Contents

Introduction

This is a new, updated and revised edition of *The Numeracy Pack*. It is intended for everyone who wants to improve their mathematics at the levels covered in the *Adult Numeracy Core Curriculum* (DfES 2001) and for anyone who wants to help someone to do so.

The core curriculum covers: number; measures, shape and space; and handling data. The pack covers selected areas of the curriculum, with special emphasis on understanding what you're doing. Curriculum references have now been included in this 4th edition where appropriate but they do not always cover all of the calculations or activities on each page. Some areas are dealt with in more detail to allow for learning and teaching.

The pack is divided into four books:

Book 1

- Introduction
- Numbers – how they're written, what they mean, how they're organised

Book 2

- The 'four rules of number' – addition, subtraction, multiplication and division, with whole numbers

Book 3

- Fractions, decimals, percentages, ratio and proportion

Book 4

- Measurement – of length, weight, capacity, time, temperature and money; handling data; using a calculator and other technology
- Answers

You don't have to start with Book 1 and work through them all. Instead, you can concentrate on the areas you need in any of the books.

The pack will help you to become more numerate, that is 'to use mathematics at a level necessary to function at work and in society in general, i.e. to understand and use mathematical information; calculate and manipulate mathematical information; interpret results and communicate mathematical information.'[1]

---

1. DfES (2001) *Adult Numeracy Core Curriculum*, London: DfES, p3.

The emphasis is on maximising your learning capacity and becoming a confident independent learner, able to tackle anything to do with numbers that you're likely to come across in everyday life and work. There are lots of opportunities to check your progress, and help with checking your work.

There is often more than one way of getting the right answer, so we've included different methods, including informal ones you can do in your head.

Sometimes it's more appropriate to get a rough idea, rather than working everything out accurately, so the pack includes help with guessing and estimation.

Some people think that maths should always be done alone and that if you ask your neighbour for help, that's cheating. We disagree. Instead, we agree with the group of adult numeracy students in Tower Hamlets, London, working with Diana Coben, who produced the poster reproduced on the back of the pack binder. After a lively discussion, the students concluded that:

Working together isn't cheating.

Using a calculator isn't cheating.

Finding out the answer from the back of the book and working out how they got it isn't cheating.

Cheating is pretending you understand when you don't.

That's when you're cheating yourself.

What do you think?

We hope you will feel free to do the maths in this pack in your head, on paper, with a calculator, computer or other calculation aid, with a friend, with a teacher, in a group or on your own – however you like.

Diana Coben and Sandy Black

Your maths life history

It may help to start your journey to becoming a confident independent learner by looking back over your past and present experiences with maths – your 'maths life history'.

Note down your answers to these ten questions in relation to the maths in your life, or use them to interview a friend, who will then interview you. You can use your answers as the basis for writing your own maths life history.

1. How would you describe your feelings about mathematics? In what circumstances do these feelings arise?

2. Describe one thing that you enjoy and one thing that you hate that involve maths.

3. Thinking back over your past life, what events involving maths stand out in your memory?

4. How would you describe your school experiences of maths?

5. In what circumstances do you work out something mathematical . . .
    i.   in your head?
    ii. on paper?
    iii. with a calculator?
    iv. another way . . . how?

6. When and how do you calculate accurately . . .
    i.   using whole numbers?
    ii. using fractions, including decimal fractions?

7. When and how do you estimate rather than calculate accurately?

8. Describe situations in which you measure each of the following: length, weight, capacity.

9. What activities do you engage in which involve an appreciation of, or the manipulation of, shape and spatial relationships?

10. What effect do you think your experience of maths has had on your opportunities in life?[2]

---

2. Coben, D. & Thumpston, G. (1996) 'Common sense, good sense and invisible mathematics', in T. Kjærgård, A. Kvamme, N. Lindén (Eds.), *Numeracy, Race, Gender and Class*, Landås, Norway: Caspar Publishing Company, p298.

When adults from all walks of life were asked these questions, these are some of the themes that emerged.

- The *brick wall* – the point (usually in childhood) at which mathematics stopped making sense; for some people it was long division, for others, fractions or algebra, while others never hit the brick wall. For those who did, the impact was often traumatic and long-lasting.

- The *significant other* – someone perceived as a major influence on the person's maths life history. The influence might be positive or negative, past or present. Significant others included, for example, a parent who tried to help with maths homework, a teacher who made the person feel stupid, a partner who undermined the person's confidence in their mathematical abilities.

- The *door* – marked 'Mathematics', locked or unlocked, which people have to go through to enter or get on in a chosen line of work or study.

- *Invisible maths* – the mathematics someone can do, but which they may not think of as maths at all, 'just common sense.'[3]

The following comments were made by people interviewed in another study, which looked into adults' use of maths in daily life.[4]

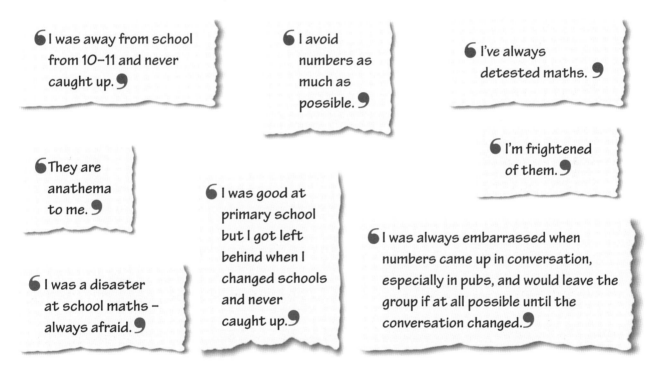

    I was away from school from 10–11 and never caught up.

    I avoid numbers as much as possible.

    I've always detested maths.

    They are anathema to me.

    I'm frightened of them.

    I was good at primary school but I got left behind when I changed schools and never caught up.

    I was a disaster at school maths – always afraid.

    I was always embarrassed when numbers came up in conversation, especially in pubs, and would leave the group if at all possible until the conversation changed.

Do any of these ring a bell with you? If so, you're not alone, but if your feelings about maths are negative you don't need to go on feeling like that. One way of recording your progress and coming to terms with your feelings about maths is to keep a maths diary.

---

3. Adapted from Coben, D. & Thumpston, G. (1996) p288.

4. Sewell, B. (1981) *Use of Mathematics by Adults in Daily Life,* Advisory Council for Adult and Continuing Education (ACACE) Enquiry Officer's Report, Leicester: NIACE.

Keeping a maths diary

If you found it helpful writing your maths life history, you might like to start a maths diary in which you record your thoughts and feelings about the maths in your life and record your progress in becoming a confident independent learner. As you go along, you may find that sometimes you're making a lot of progress, then at other times you seem to be getting nowhere, even going backwards, forgetting some of the things you thought you'd learned. Learning is like this – sometimes you go 'two steps forward and three steps back' and it can be very discouraging. Stick at it. Your writing about maths can help you to identify your strengths and weaknesses. Build on your strengths and work on your weak points. If you're learning in a group, talk to other people about their methods for doing maths, and how they feel about their own maths. Allow yourself plenty of time to work something out if you know you find it difficult. Take deep breaths and use relaxation techniques if you find yourself getting anxious. Prove to yourself and others you can do it.

This is an entry from one adult numeracy student's maths diary:

> ❝I did maths today. It is opening up a new world for me as I always believed I was stupid . . . today my tutor taught me how to identify a twenty-four hour clock . . . I am also beginning to use a calculator and lose my fear of it. Each time I go I come away feeling I want the class to carry on all day. Maybe one day I'll know enough to sit my GCSE in maths.[5]❞

May a new world open up for you too.

---

5. Quoted in Coare, P. and A. Thomson (Eds.) (1996) *Through the Joy of Learning: Diary of 1,000 adult learners,* Leicester: NIACE.

You can count in your head or use your fingers, an abacus or other mechanical aid, or make marks for counting.

There are many different ways of counting using the fingers. If you count one to a finger, you get five for each hand and ten for both hands.

An Indian method counts the segments of the finger between the joints and a Chinese method enables people to count up to thousands using their fingers.

Tallying

Making marks like this |||| is called tallying and the individual marks are called tallies.

It may help you to keep count if you group the tallies together in groups of five. A common way of doing this is to write |||| and then cross through the tallies with the fifth line, like this: ‖‖. That way you'll know you've got five in each group.

Tallying is used when collecting and collating data, for example, to count the numbers of vehicles passing a certain point on a road, or the number of people getting on a bus – in any situation where it's quicker and more reliable to keep a tally than to try to remember numbers in your head.

Tallying is an ancient way of keeping count, and there are many ways of doing it, for example, making marks on stone or pieces of wood or making groups of pebbles or any other readily available items.

General knowledge of whole numbers

The numbers we use for counting, **0, 1, 2, 3, 4, 5, 6, 7, 8, 9, 10, 11, 12**... and so on, are called whole numbers. They are also called natural numbers or positive integers.

They start with zero (0) and go on forever, to infinity. There is no last whole number.

Fill in the gaps:

1. There are _____ days in a week.

2. There are _____ days in a year.

3. Twice two are _____ .

4. _____ is one more than six.

5. Add one to three and you get _____ .

6. Take two away from six and you get _____ .

7. _____ is the same as half a dozen.

8. _____ makes a pair.

9. _____ is one less than seven.

10. _____ is one more than nine.

11. Three is half of_____ .

12. Eight is half of_____ .

How many?

How many people are there in the room?

How many are women?

How many are men?

How many tables are there in the room?

How many chairs are there?

How many windows are there in the room?

How many doors are there?

How many pens are there on your table?

How many coins have you got in your pocket or bag?

How many of them are 50p pieces?

How many of them are 10p pieces?

How many days until next Friday?

How many months since your last birthday?

What if?

What if two more people came into the room, how many would there be then?

What if you started with 50p and then lost two 10p pieces – how much would you have left?

N1/E1.1, MSS1/E1.1, MSS1/E1.2

How many shapes?

What are these shapes called?

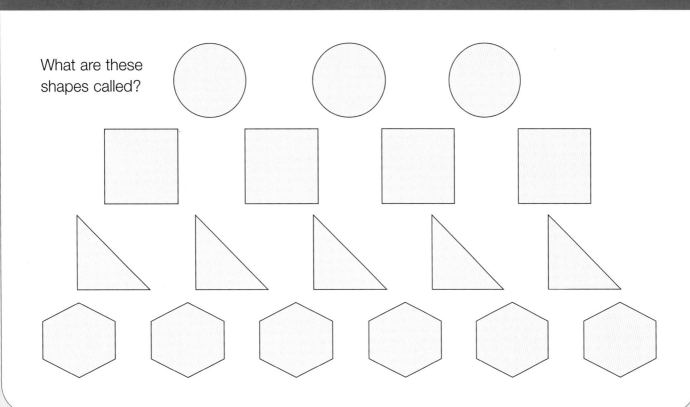

MSS2/E1.1, MSS2/E2.1

Count the dots	Write the number	Write the word
	0	*zero*
●		
● ●		
● ● ●		
● ● / ● ●		
● ● / ● / ● ●		
● ● / ● ● / ● ●		
● ● / ● ● ● / ● ●		
● ● / ●●● ●● / ● ●		
● ● / ● ● ● / ● ● ●		
●● ●● / ●● ●● / ●●● ●●●		

N1/E1.2

N1/E1.2

0 1 2 3 4 5 6 7 8 9

These are the numbers that we use.

They are known as Arabic or Indo-Arabic numbers because they were originally developed in India and then spread throughout the Arab world.

There are only ten of them but we can make any number by putting them together in different ways.

The system that we use to do this is called 'place value' because the numbers have a value depending on where they are placed. For example, in the number 21 the figure 2 means two tens and the figure 1 means one unit (unit is another name for one).

In the number 635 the figure 6 means six hundred (H); the figure 3 means three tens (T); and the figure 5 means five units (U).

To help you remember, you can put the letters H T U above the figures in the number.

H T U
6 3 5

The individual figures which go to make up a number are also called digits, after the Latin word for finger.

Place value is vitally important in arithmetic. If you put a figure in the wrong place – if you write 19 instead of 91 for instance, it makes a lot of difference.

The place value system we use is based on ten (it's called a base 10 system). Any figure in the tens column is worth ten times what it would be worth in the units column. Similarly, any figure in the hundreds column is worth ten times what it would be worth in the tens column and so on.

There is no end to the system, you could go on writing the same figure, for instance 4, again and again, and each 4 would be worth ten times more than its neighbour to the right, and ten times less than its neighbour to the left.

4444444444444444444444444444444444444444444444444444444444

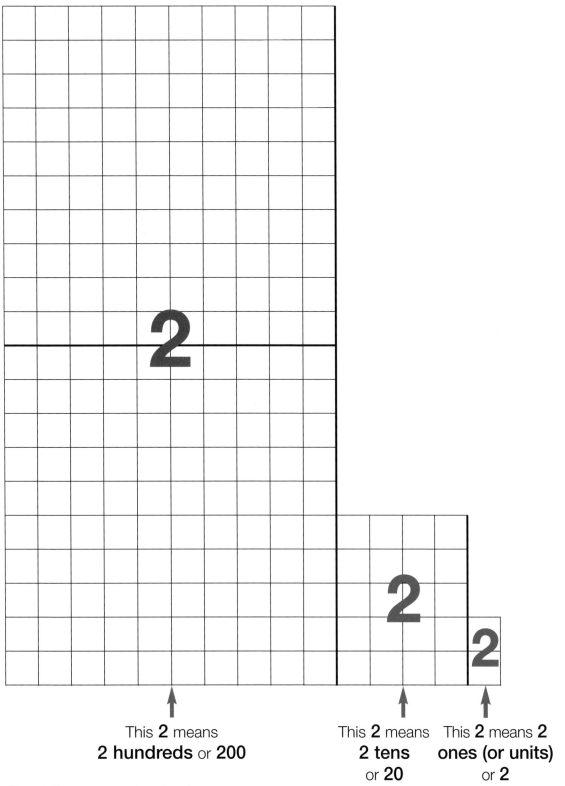

This **2** means
2 hundreds or **200**

This **2** means
2 tens
or **20**

This **2** means **2**
ones (or units)
or **2**

Count the squares to check.

222 = **2** hundreds + **2** tens + **2** units.

We say: **two hundred and twenty-two**.

N1/E3.1

Put one number in each column to show the value of that number.

For instance, 2 in the 100s column has the value 200.

2 in the units column has the value 2.

The bigger the number the more columns it uses.

$$10 \times 100 = 1000 \qquad 10 \times 10 = 100 \qquad 10 \times 1 = 10$$

1000s THOUSANDS	100s HUNDREDS	10s TENS	1s UNITS

Place value

	Thousands Th	Hundreds H	Tens T	Units U	WRITE THE NUMBER IN WORDS
5432	5	4	3	2	five thousand, four hundred and thirty-two
7628					
594					
2021					
3603					
4670					
67					
6042					

N1/E3.1, N1/L1.1

Roman numerals

Roman numbers do not use place value. These are the Roman numerals equivalent to the Indo-Arabic numbers 1 to 10.

I   II   III   IV   V   VI   VII   VIII   IX   X

The Romans based these numbers on the fingers and hands:

I

II   **one, two and three fingers**

III

V is a whole hand (five fingers) with the fingers together and the thumb stretched out to make a **V** shape. **IV** means one before five (4), and **VI** means one after five (6). **VII** means seven, and **VIII** means eight.

X is both hands (ten fingers) with fingers and thumbs together and the hands straight, held together to make a cross. **IX** means one before ten (9) and **XI** means one after ten (11) and so on.

The larger numbers relate to the names of the numbers in Latin.

L 50   C 100   D 500   M 1000

The principle is the same for all Roman numbers. A letter to the left means it is taken away from its neighbour to the right and a letter to the right means it is added on to its neighbour to the left. For example, MCMXCVI is 1996 because MCM means 2 thousand less one hundred (1900), XC means one hundred less ten (90) and VI means one more than five (6). To find the date you add these together: 1900 + 90 + 6 = 1996.

Years in the twenty-first century are simpler to write in Roman numerals: MM is 2000, MMIV is 2004, and so on.

Nowadays, Roman numerals are mainly used on clock faces, after the names of kings and queens and in film credits, to show the year the film was made. They are generally thought to look formal and impressive and are therefore used whenever such an effect is intended.

What would these Roman numbers be, written as Indo-Arabic numbers?

XV   XX   DC   VIII   XIX

XXIII   XC   LX   MD

Odd and even numbers

The numbers 1, 2, 3, 4, 5, 6, 7, 8, 9, 10, 11, 12, 13, 14 . . . and so on, make a pattern of odd and even numbers.

1 is an odd number, 2 is an even number, 3 is odd, 4 is even and so on. Any whole number is either even or odd.

Even numbers can be divided by 2.

Odd numbers cannot be divided equally by 2, for example, 3, 17 or 21.

Any whole number ending in an even number (0, 2, 4, 6 or 8) is an even number.

Any whole number ending in 1, 3, 5, 7 or 9 is an odd number, for example, 539.

If you multiply two even numbers together the answer will always be an even number.

If you multiply two odd numbers together the answer will always be an odd number.

Put a circle round every odd number:

| 5 | 12 | 24 | 7 | 11 | 18 | 19 | 22 | 25 |

Put a circle round every even number:

| 2 | 9 | 24 | 8 | 30 | 15 | 29 |

Recognising numbers

1. Draw a line under all the numbers that come after 3:

| 7 | 2 | 5 | 1 | 6 | 11 | 76 | 10 |

2. Put these numbers into order with the smallest number first (this is called ascending order):

| 8 | 6 | 9 | 1 | 3 | 7 | 5 | 10 | 4 | 2 |

3. Put these numbers into reverse order, with the largest number first (this is called descending order):

| 19 | 50 | 14 | 7 | 5 | 24 | 37 | 95 | 62 |

Writing the numbers 1 to 20

1	one
2	two
3	three
4	four
5	five
6	six
7	seven
8	eight
9	nine
10	ten
11	eleven
12	twelve
13	thirteen
14	fourteen
15	fifteen
16	sixteen
17	seventeen
18	eighteen
19	nineteen
20	twenty

Writing the numbers 21 to 100

21	twenty-one
22	twenty-two
23	twenty-three
24	twenty-four
25	twenty-five
26	twenty-six
27	twenty-seven
28	twenty-eight
29	twenty-nine
30	thirty
31	thirty-one
32	thirty-two
33	thirty-three
34	thirty-four
35	thirty-five
36	thirty-six
37	thirty-seven
38	thirty-eight
39	thirty-nine
40	forty
50	fifty
60	sixty
70	seventy
80	eighty
90	ninety
100	one hundred

Writing large numbers

The figures in large numbers are grouped together in threes, starting from the right hand end, in order to make them easier to read. Each group of three figures may be separated either with a comma or, sometimes, a small gap. For example, one million may be written as 1,000,000 or 1 000 000.

Put commas into the following figures where appropriate:

100	one hundred
110	one hundred and ten
120	one hundred and twenty
1000	one thousand
1100	one thousand one hundred (or eleven hundred)
1110	one thousand one hundred and ten
10000	ten thousand
20000	twenty thousand
100000	one hundred thousand
110000	one hundred and ten thousand
100111	one hundred thousand one hundred and eleven
1000000	one million

Look at the number below:

1,264,509

We say: one million, two hundred and sixty-four thousand, five hundred and nine.

This means: 1 million; 2 hundred thousands; 6 tens of thousands; 4 thousands; 5 hundreds; no tens and nine ones.

Note that in the USA, 1,000,000,000 is called one billion. A British billion is traditionally one million million: 1,000,000,000,000, which is one thousand times larger than a US billion. In the UK, 1,000,000,000 is traditionally called one thousand million. So if the term billion is used, you need to know whether it refers to a US billion or a British billion.

Say these numbers:

405	23,678	6,987,453	1,094	286	42,900

More than and less than

This symbol means more than or greater than:

>

This symbol means less than:

<

They're called inequality symbols and they're used for comparing numbers or quantities.

For example, 19 is more than 6, so you write that as 19 > 6.

Fill in the missing > or < symbols in these pairs of numbers:

4   16

88   9

23   90

34   23

13   28

2   7

The other inequality symbols are:

≥   greater than or equal to

≤   less than or equal to

The equals sign is =

Counting in 2s – even numbers

2	two
4	four
6	six
8	eight
10	ten
12	twelve
14	fourteen
16	sixteen
18	eighteen
20	twenty
22	twenty-two
24	twenty-four
26	twenty-six
28	twenty-eight
30	thirty
32	thirty-two
34	thirty-four
36	thirty-six
38	thirty-eight
40	forty

	Counting in 5s			Counting in 10s	
5	five		**10**	ten	
10	ten		**20**	twenty	
15	fifteen		**30**	thirty	
20	twenty		**40**	forty	
25	twenty-five		**50**	fifty	
30	thirty		**60**	sixty	
35	thirty-five		**70**	seventy	
40	forty		**80**	eighty	
45	forty-five		**90**	ninety	
50	fifty		**100**	one hundred	
55	fifty-five		**110**	one hundred and ten	
60	sixty		**120**	one hundred and twenty	
65	sixty-five		**130**	one hundred and thirty	
70	seventy		**140**	one hundred and forty	
75	seventy-five		**150**	one hundred and fifty	
80	eighty		**160**	one hundred and sixty	
85	eighty-five		**170**	one hundred and seventy	
90	ninety		**180**	one hundred and eighty	
95	ninety-five		**190**	one hundred and ninety	
100	one hundred		**200**	two hundred	

Ordinal numbers

Ordinal numbers are commonly used in dates, for example, 7th March, and for birthdays and anniversaries, to refer to floors in a building (2nd floor, 3rd floor) and in races and other competitions. Think of some other examples.

1st	first	20th	twentieth
2nd	second	21st	twenty-first
3rd	third	22nd	twenty-second
4th	fourth	23rd	twenty-third
5th	fifth	24th	twenty-fourth
6th	sixth	25th	twenty-fifth
7th	seventh	26th	twenty-sixth
8th	eighth	27th	twenty-seventh
9th	ninth	28th	twenty-eighth
10th	tenth	29th	twenty-ninth
11th	eleventh	30th	thirtieth
12th	twelfth	31st	thirty-first
13th	thirteenth	40th	fortieth
14th	fourteenth	50th	fiftieth
15th	fifteenth	60th	sixtieth
16th	sixteenth	70th	seventieth
17th	seventeenth	80th	eightieth
18th	eighteenth	90th	ninetieth
19th	nineteenth	100th	one hundredth

The sign	It means
+	plus add and positive
—	minus take away subtract less negative
×	multiply times of
÷ or /	divide share . . . out of . . .
=	equals …is the same as… …is equivalent to…

The signs **+ — ×** and **÷** tell you what operation to do in a calculation.

Also, the plus sign before a number indicates a positive number, for example, $+4$ is **plus 4**.

The minus sign before a number indicates a negative number, for example, -4 is **minus 4** (see 'Positive and negative numbers' in this book).

A statement with an equals sign is called an equation. The values either side of the equals sign must balance. For example, $2 + 2 = 4$ balances because the value of $2 + 2$ is the same as the value of **4**.

Addition, subtraction, multiplication and division are called the four rules or sometimes the four operations in arithmetic.

N1/E1.6, N1/E2.7, N1/E3.9

See the difference a sign makes:

$$4 + 2 = 6 \qquad\qquad 12 + 3 = 15$$

$$4 - 2 = 2 \qquad\qquad 12 - 3 = 9$$

$$4 \times 2 = 8 \qquad\qquad 12 \times 3 = 36$$

$$4 \div 2 = 2 \qquad\qquad 12 \div 3 = 4$$

Try these

Fill in the numbers:

$$20 + 4 =$$

$$20 - 4 =$$

$$20 \times 4 =$$

$$20 \div 4 =$$

Fill in the signs and the numbers:

$$9 \quad 3 = \qquad\qquad 10 \quad 5 =$$

$$9 \quad 3 = \qquad\qquad 10 \quad 5 =$$

$$9 \quad 3 = \qquad\qquad 10 \quad 5 =$$

$$9 \quad 3 = \qquad\qquad 10 \quad 5 =$$

There are different ways of saying and writing calculations. They can be written along a line or with numbers above one another and the sign on either the right or left, as shown:

Add $6 + 7 = 13$

$$\begin{array}{r} 6\,+ \\ 7 \\ \hline 13 \end{array}$$

$$\begin{array}{r} 6 \\ +\,7 \\ \hline 13 \end{array}$$

It says: **6** add **7** equals **13**

or   **6** plus **7** equals **13**

or   **6** added to **7** equals **13**

or   **6** and **7** equals **13**

or   **6** and **7** makes **13**

or   **6** and **7** is **13**

or   **6** and **7** are **13**

Take away $10 - 6 = 4$

$$\begin{array}{r} 10\,- \\ 6 \\ \hline 4 \end{array}$$

$$\begin{array}{r} 10 \\ -\,6 \\ \hline 4 \end{array}$$

It says: **10** take away **6** equals **4**

or   **10** take away **6** leaves **4**

or   **10** minus **6** equals **4**

or   **10** minus **6** leaves **4**

or   **10** subtract **6** equals **4**

or   **10** subtract **6** leaves **4**

You can say the smaller number first: **6** from **10** leaves **4**

or   **6** taken from **10** leaves **4**

or   **6** subtracted from **10** leaves **4**

N1/E1.6, N1/E2.7

Multiply $3 \times 4 = 12$

$$\begin{array}{r} 3 \times \\ 4 \\ \hline 12 \end{array} \qquad \begin{array}{r} 3 \\ \times 4 \\ \hline 12 \end{array}$$

It says: **3** multiplied by **4** equals **12**

or   **3** times **4** equals **12**

or   **3** times **4** is **12**

or   **3** times **4** makes **12**

or   **3** fours are **12**

or   **3** fours make **12**

Divide $15 \div 3 = 5$

$$\frac{15}{3} = 5 \qquad 3 \overline{)\,15}^{\,5}$$

It says: **15** divided by **3** equals **5**

or   **15** divided by **3** is **5**

Or alternatively: $3 \overline{)\,15}^{\,5}$ **3** divided into **15** equals **5**

or   **3** into **15** equals **5**

or   **3** into **15** goes **5** times

or   **3** into **15** is **5**

Write these calculations with signs:

6 and 6 makes 12   7 plus 6 makes 13

15 take away 5 leaves 10   6 into 18 goes 3 times

18 divided by 6 equals 3   fours into 20 is 5

9 minus 4 equals 5   5 multiplied by 6 makes 30

3 times 5 equals 15   5 from 10 leaves 5

5 fives are 25   9 from 16 leaves 7

Take 5 counters and add another 4 to them.

How many in all?

Write an **add** sum to show what you have done.

$$5 + 4 = 9$$ five add four makes nine

You should get the same answer if you start with 4 and add 5. Try it.

Now start with 9 counters and take away 5.

How many are left?

You can write a **take away** or **subtract** sum to show this.

$$9 - 5 = 4$$ nine take away five leaves four

Start again with 9 counters and this time take away 4.

How many are left?

You can write this as:

$$9 - 4 = 5$$ nine take away four leaves five

So adding and taking away go together like this:

$$5 + 4 = 9 \qquad 9 - 5 = 4 \qquad 9 - 4 = 5$$

N1/M8.7, N1/M8.8, N1/M8.10
N1/E1.4, N1/E1.5, N1/E2.4

In the same way, take any other groups of counters.

Take 12 and add 7 to make 19.

$$12 + 7 = 19$$

Start with 19 and take away 12.

How many are left?

$$19 - 12 = 7$$

Start with 19 and take away 7.

How many are left?

$$19 - 7 = 12$$

Look at these and see how add and take away relate to each other:

$2 + 3 = 5$	$15 + 10 = 25$	$8 + 9 = 17$
$5 - 2 = 3$	$25 - 10 = 15$	$17 - 9 = 8$
$5 - 3 = 2$	$25 - 15 = 10$	$17 - 8 = 9$

Complete these:

$8 + 3 = 11$	$14 + 4 = 18$	$10 + 11 = 21$	$16 + 8 = 24$
$11 - 8 =$	$18 - 4 =$	$21 - 10 =$	$24 - 8 =$
$11 - 3 =$	$18 - 14 =$	$21 - 11 =$	$24 - 16 =$

Write two take away sums from each of these add sums. Use counters to help.

$4 + 6$	$12 + 9$	$13 + 15$	$3 + 8$

You can use adding and taking away to check your calculations. For example, if you think that $5 + 2 = 7$, you can check by trying $7 - 2$ and $7 - 5$. If you get 5 for the first one and 2 for the second one you can be fairly sure you're right.

Write these sentences using figures, and the signs + − × ÷ = instead of words

1. Four plus two makes six

2. Twenty-three minus seven leaves sixteen

3. Eighteen added to thirty makes forty-eight

4. One hundred and forty-four equals twelve times twelve

5. Three eights are twenty-four

6. Thirty-six divided by nine is four

7. Seventeen plus six take away three makes twenty

Write these in words

$28 \div 4 = 7$

$150 - 25 = 125$

$327 + 16 = 343$

$9 \times 18 = 162$

$54 + 6 - 3 = 57$

$12 \times 12 = 144$

$6 = 2 \times 3$

$2 \times 4 \times 6 = 48$

$120 = 12 \times 10$

Write the name of the sign in each of these

8 + 4

16 ÷ 8

3 × 6

44 ÷ 11

18 − 6

17 + 22

13 − 5

9 × 7

16 + 27

36 − 20

14 ÷ 2

16 − 7

2 + 19

33 + 8

Write the sign + − × ÷ next to each of these

fifteen **divided** by three

eight **plus** twelve

24 **divided** by 6

12 **times** 9

4 **multiplied** by 5

21 **take away** 13

102 **added** to 316

56 **subtracted** from 100

12 **plus** 8 **plus** 16

20 **times** 6

15 **add** 36

2 **times** 2 **times** 2

72 **minus** 24

18 **divided** by 2

100 **shared** into 10 parts

Add or multiply?

+ plus or add   × multiply

Put in the missing sign
example: 5 + 9 = 14

3   5 = 15

2   6 = 8

9   2 = 18

3   4 = 12

7   2 = 14

16   4 = 20

2   10 = 20

4   6 = 24

2   8 = 16

7   4 = 28

7   8 = 56

6   5 = 30

48   2 = 50

3   7 = 21

9   3 = 27

6   7 = 42

9   9 = 81

5   6 = 11

3   3 = 9

Put in the missing signs
+ − × ÷

example: 3 × 3 = 9

6   7 = 13

18   2 = 9

100   10 = 10

17   5 = 12

9   9 = 81

21   8 = 13

4   5 = 20

158   16 = 142

15   4 = 60

120   60 = 60

24   6 = 4

12   3 = 36

16   4 = 4

13   25 = 38

125   25 = 5

6   12   3 = 21

10   5   3 = 12

2   2   2 = 8

Which signs?

Put a ring round the correct sum to match the words on the left-hand side.

1. The number of days in two weeks.

   $2 + 7$ $7 - 2$
   2×7 $7 + 7$
   $7 \div 2$ $2 + 2$

2. Jim is 9 today.
   How old will he be in 8 years time?

   9×8 $9 + 9$
   $9 \div 8$ $9 + 8$
   $9 - 8$ $8 + 8$

3. My newspaper costs 45p a day.
   How much is this for 5 days?

   5×45 45×5
   $45 + 5$ $45 - 5$
   $45 \div 5$ $5 \div 45$

4. We are walking on a 23-mile hike to Wike.
   The signpost says 'WIKE 16 miles'.
   How far have we travelled?

   16×23 23×16
   $23 + 16$ $23 \div 16$
   $16 - 23$ $23 - 16$

5. A block of flats has 14 floors.
   There are 84 flats altogether.
   How many on each floor?

   $84 + 14$ $84 \div 14$
   $14 - 84$ $84 - 14$
   84×14 $14 \div 84$

6. David's uncle is 41.
   He is 22 years older than David.
   How old is David?

   22×41 41×22
   $41 - 19$ $41 + 22$
   $41 - 22$ $41 \div 22$

7. Aisha has 24 CDs in her collection.
   Last week she lost 5 of them.
   How many has she left?

   $5 + 24$ 24×5
   $24 - 5$ $24 \div 5$
   $29 - 5$ $5 - 24$

8. There are 6 squares in a bar of chocolate.
   How many squares in 8 bars?

   $8 + 6$ $14 - 6$
   $8 \div 6$ $6 \div 8$
   14×6 6×8

The number square

The number square can be used in a variety of ways, as a counting square, as a multiplication square and as an aid to understanding factors.

As a **counting square:** fill in the numbers 1 to 10 from left to right along the top line of squares. Carry on from left to right in the next row, and so on until the square is full of numbers. You could then shade every second number or every fifth and so on as a preparation for multiplication.

As a **multiplication square:** fill in the numbers 1 to 10 from left to right along the top line of squares. Now write the numbers 2 to 10 down the left hand side of the square. Then work along one line at a time, saying $2 \times 1 = 2$, $2 \times 2 = 4$, $2 \times 3 = 6$ and so on, as shown below.

Work through each line until the square is full. You will find you've got a handy square on which to check your tables. You may also find that you know more of your tables than you thought.

Factors: look at the multiplication square. Pick a number and see how many times you can find the same number in the square.

For example, $36 = 6 \times 6$ $36 = 4 \times 9$ $36 = 9 \times 4$

There is more about factors on p38 in this book.

1	2	3	4	5	6	7	8	9	10
2	4	6	8	10	12				
3	6	9	12						
4	8	12							
5	10	15							
6	12								
7	14								
8	16								
9	18								
10	20								

				34					

4 groups of **2**

How many in all?

We write this   **2 + 2 + 2 + 2 = 8** (add)

or this   **4 × 2 = 8** (multiply)

Multiplying is the same as adding equal groups.

5 groups of **3**

How many in all?

We write this   **5 × 3 = 15**

or this   **3 + 3 + 3 + 3 + 3 = 15**

Multiplying is quicker, and simpler to write.

Write add and multiply sums for these:

         5 groups of **4**

          6 groups of **2**

N1/E2.5, N1/E2.2, N1/E2.7

Write the multiplication sums that match the addition sums on the left-hand side

ADD +	MULTIPLY ×	ADD +	MULTIPLY ×
3 + 3 3 ――― 9	3 × 3 ――	6 + 6 ――	
4 + 4 4 4 ――		7 + 7 7 ――	
5 + 5 5 ――		10 + 10 10 10 ――	

N1/E2.2, N1/E2.3, N1/E2.5

Add and multiply using counters

You have 15 counters. Share them into 3 equal groups.

How many in each group? Draw the groups.

Write an addition (+) sum to show how you have divided the 15 counters.

Write a multiplication (×) sum which shows the same thing.

Now divide the 15 counters into equal groups of a different size. How many in each group?

Draw the groups. Write addition (+) and multiplication (×) sums to show how you have divided the counters this time.

Starting with 20 counters, find 2 different ways of sharing them into groups of equal size.

For each way, draw the groups, and write an addition (+) and a multiplication (×) sum to show what you have done.

Starting with 24 counters, find 3 different ways of sharing them into groups of equal size.

For each way, draw the groups, and write an addition (+) sum and a multiplication (×) sum to show what you have done.

Take 6 counters, double them, double again.

How many have you got now? Write down the calculation you've done.

N1/E2.5, N1/E3.5, N1/E3.6, N1/E3.9

Imagine a bar of chocolate like this.
There are **12** pieces altogether, but we can break it up in different ways.

If we break it this way we have 4 lots of 3 pieces.

4 × 3 = 12

If we break it this way, we have 3 lots of 4 pieces.

3 × 4 = 12

We have turned the sum around, but still have the same number of pieces. Now try the same thing with these groups of rectangles: group them in two different ways as indicated. Write the sum underneath the groups.

4 lots of 2

2 lots of 4

5 lots of 3

3 lots of 5

4 lots of 5

5 lots of 4

N1/E2.5

What do we know about the number 12?

We know **3 × 4 = 12**   and   **12 ÷ 4 = 3**

12 ÷ 3 = 4

The numbers **3** and **4** are called factors of **12** because they divide into **12** a whole number of times.

We also know **2 × 6 = 12**   and **12 ÷ 6 = 2**

12 ÷ 2 = 6

The numbers **2** and **6** are also factors of **12**.

How many ways are there of splitting **12** into factors?

Look at the number of ways of drawing a block of 12 squares.

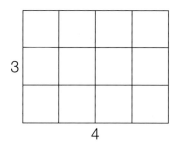

12 = 3 × 4 or

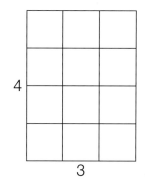

12 = 4 × 3

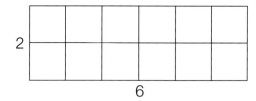

12 = 2 × 6 or

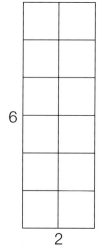

12 = 6 × 2

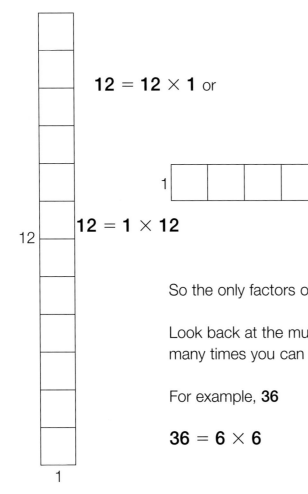

12 = 12 × 1 or

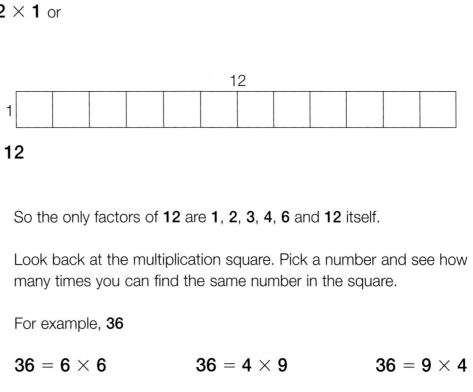

12 = 1 × 12

So the only factors of **12** are **1**, **2**, **3**, **4**, **6** and **12** itself.

Look back at the multiplication square. Pick a number and see how many times you can find the same number in the square.

For example, **36**

36 = 6 × 6   **36 = 4 × 9**   **36 = 9 × 4**

Prime factors are factors of a number which are also prime numbers. A prime number only has two factors, itself and one.

Take one set of factors, for example: 4 × 9 and find the factors of each number.

In this case   **4 = 2 × 2**   and   **9 = 3 × 3**

so you could write   **36 = 2 × 2 × 3 × 3**

so **2** and **3** are prime factors of **36**

Factorise these numbers:

48   **33**   **100**   **24**   **16**   **39**

What do you notice about these factors?

There's more on factors and factorising in Book 2.

×	+
1 × 3 = 3	3
2 × 3 = 6	3 + 3 = 6
3 × 3 = 9	3 + 3 + 3 = 9
4 × 3 = 12	3 + 3 + 3 + 3 = 12
5 × 3 = 15	3 + 3 + 3 + 3 + 3 = 15
6 × 3 = 18	3 + 3 + 3 + 3 + 3 + 3 = 18
7 × 3 = 21	3 + 3 + 3 + 3 + 3 + 3 + 3 = 21
8 × 3 = 24	3 + 3 + 3 + 3 + 3 + 3 + 3 + 3 = 24
9 × 3 = 27	3 + 3 + 3 + 3 + 3 + 3 + 3 + 3 + 3 = 27
10 × 3 = 30	3 + 3 + 3 + 3 + 3 + 3 + 3 + 3 + 3 + 3 = 30
11 × 3 = 33	3 + 3 + 3 + 3 + 3 + 3 + 3 + 3 + 3 + 3 + 3 = 33
12 × 3 = 36	3 + 3 + 3 + 3 + 3 + 3 + 3 + 3 + 3 + 3 + 3 + 3 = 36

Write these as multiplication sums.

1.   5 + 5 + 5 + 5 = 20

2.   7 + 7 + 7 + 7 + 7 = 35

3.   8 + 8 + 8 = 24

4.   9 + 9 + 9 + 9 + 9 + 9 + 9 = 63

5.   2 + 2 + 2 + 2 + 2 + 2 = 12

6.   6 + 6 = 12

7.   4 + 4 + 4 + 4 = 16

8.   12 + 12 + 12 + 12 + 12 + 12 = 72

9.   three sixes are eighteen

10.   five fours are twenty

11.   nine eights are seventy-two

12.   fifteen times ten makes a hundred and fifty

Multiply and divide

When we put together equal groups we multiply them.

3 groups of **5** make **15**

3 × 5 = 15

Three times five equals fifteen

When we split something into equal parts, we divide it.

When **15** is divided into **3** equal groups, there are **5** in each group.

So: **15 ÷ 3 = 5**

15 divided by **3** equals **5**

When 15 is divided into 5 equal parts, there are 3 in each group.

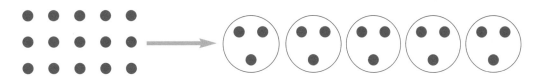

So: **15 ÷ 5 = 3**

15 divided by **5** equals **3**

Multiplying and dividing go together like this:

5 × 3 = 15	**15 ÷ 3 = 5**	or	$\begin{array}{r} 5 \\ 3\overline{)15} \end{array}$ $\begin{array}{r} 3 \\ 5\overline{)15} \end{array}$
3 × 5 = 15	**15 ÷ 5 = 3**		

Positive and negative numbers

A positive number is a number above zero. Positive numbers may be written with a plus sign in front of the number, for example, +6, +7. Normally a number without a sign in front is a positive number so if you don't see a plus sign in front, you can assume the number is positive.

A negative number is a number below zero (0). Negative numbers are always written with a minus sign in front of the number, for example, −3, −4, and so on.

Zero is the point of equilibrium between positive and negative numbers.

Look at the number line and the thermometer below:

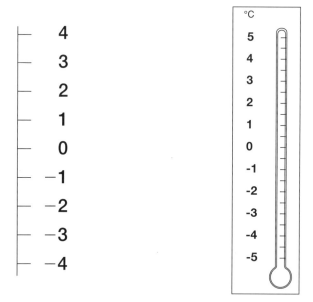

Negative numbers were developed for use in accounting and they are still used in keeping accounts. A negative number denotes an amount owed and a positive number denotes an amount in hand.

Negative numbers are also used in measuring temperature. The scale on a thermometer shows temperatures above and below zero; these are positive and negative numbers respectively.

If there is a minus sign in front of a sum of money, then the higher the number, the greater the amount owed.

On a thermometer, the higher the negative number the lower the temperature (for more on this, see 'Reading the thermometer' in Book 4).

Ordering positive and negative numbers

You can arrange a list of numbers in order of size, either starting with the lowest through to the highest number, or *vice versa*.

For example, the list: 9, −3, 6, −1, 5, could be ordered either as 9, 6, 5, −1, −3 (starting with the highest number) or as −3, −1, 5, 6, 9 (starting with the lowest number).

It may help to think of the numbers positioned on the number line, or on a thermometer, on either side of zero.

N1/L2.1

Calculating with negative numbers

Adding a positive number always means going up the number line.

For example, $−3 + 1$ means starting with minus 3 and moving one place up the number line, giving you the answer $−2$.

Adding a negative number means going down the number line.

For example, $−3 + −2$ is $−5$.

It's the same as subtracting a positive number.

For example, $−3 + −2$ is the same as $−3 − 2$, which equals $−5$.

When multiplying negative numbers, start by ignoring the signs and just multiply the numbers. Then look at the signs. If both numbers are positive, then the answer will be positive; if one number is positive and the other negative, then the answer will be negative.

For example, $−4 × 2 = −8$.

Check this by doing it as a repeated addition:

$−4 + −4 = −8$ because you're moving down the number line.

Dividing negative numbers works on the same principle. Divide the numbers first, then look at the signs: if both numbers are positive, then the answer will be positive; if one number is positive and the other negative, then the answer will be negative.

For example, $−6 ÷ −2 = 3$, but $−6 ÷ 2 = −3$.

There is more on negative numbers in Book 4.

N1/L1.2, N1/L2.1

Answers

General knowledge of whole numbers, p.8

1. There are seven days in a week.
2. There are three hundred and sixty-five days in a year.
3. Twice two are four.
4. Seven is one more than six.
5. Add one to three and you get four.
6. Take two away from six and you get four.
7. Six is the same as half a dozen.
8. Two makes a pair.
9. Six is one less than seven.
10. Ten is one more than nine.
11. Three is half of six.
12. Eight is half of sixteen.

Place value, p.15

	TH	H	T	U	WRITE THE NUMBER IN WORDS
5432	5	4	3	2	Five thousand, four hundred and thirty-two
7628	7	6	2	8	Seven thousand, six hundred and twenty-eight
594		5	9	4	Five hundred and ninety-four
2021	2	0	2	1	Two thousand and twenty-one
3603	3	6	0	3	Three thousand, six hundred and three
4670	4	6	7	0	Four thousand, six hundred and seventy
67			6	7	Sixty-seven
6042	6	0	4	2	Six thousand and forty-two

(Column headers: Thousands, Hundreds, Tens, Units)

Roman numerals, p.16

XV = 15   XX = 20   DC = 600   VIII = 8

XIX = 19   XXIII = 23   XC = 90   LX = 60   MD = 1500

Odd and even numbers, p.17

The odd numbers are 5, 7, 11, 19, 25

The even numbers are 2, 24, 8, 30

Recognising numbers, p.17

1. The numbers after 3 are 7, 5, 6, 11, 76, 10
2. In ascending order the numbers are 1, 2, 3, 4, 5, 6, 7, 8, 9, 10
3. In descending order the numbers are 95, 62, 50, 37, 24, 19, 14, 7, 5

More than and less than, p.20

The missing inequality symbols in these pairs of numbers are $4 < 16$, $88 > 9$, $23 < 90$, $34 > 23$, $13 < 28$, $2 < 7$

The signs, p.24

Fill in the numbers:

$20 + 4 = 24$
$20 - 4 = 16$
$20 \times 4 = 80$
$20 \div 4 = 5$

Fill in the signs and the numbers:

$9 + 3 = 12$ $10 + 5 = 15$
$9 - 3 = 6$ $10 - 5 = 5$
$9 \times 3 = 27$ $10 \times 5 = 50$
$9 \div 3 = 3$ $10 \div 5 = 2$

Talking about calculations, p.26

$6 + 6 = 12$ $7 + 6 = 13$
$15 - 5 = 10$ $18 \div 6 = 3$
$18 \div 6 = 3$ $20 \div 4 = 5$
$9 - 4 = 5$ $5 \times 6 = 30$
$3 \times 5 = 15$ $10 - 5 = 5$
$5 \times 5 = 25$ $16 - 9 = 7$

Add and take away, p.28

$11 - 8 = 3$; $18 - 4 = 14$; $21 - 10 = 11$; $24 - 8 = 16$;

$11 - 3 = 8$; $18 - 14 = 4$; $21 - 11 = 10$; $24 - 16 = 8$;

$4 + 6 = 10$ $10 - 4 = 6$ $10 - 6 = 4$
$12 + 9 = 21$ $21 - 9 = 12$ $21 - 9 = 12$
$13 + 15 = 28$ $28 - 13 = 15$ $28 - 15 = 13$
$3 + 8 = 11$ $11 - 3 = 8$ $11 - 8 = 3$

Write these sentences using figures, and the signs + − × ÷ instead of words, p.29

1. $4 + 2 = 6$
2. $23 − 7 = 16$
3. $18 + 30 = 48$
4. $144 = 12 × 12$
5. $3 × 8 = 24$
6. $36 ÷ 9 = 4$
7. $17 + 6 − 3 = 20$

Write these in words, p.29

(It is possible to have different correct answers.)

Twenty-eight divided by four is seven
Twelve times twelve equals one hundred and forty-four
One hundred and fifty minus twenty-five is one hundred and twenty-five
Six equals two multiplied by three (or two times three)
Three hundred and twenty-seven plus sixteen equals three hundred and forty-three
Two times four times six equals forty-eight
Nine multiplied by eighteen is one hundred and sixty-two
One hundred and twenty equals twelve times ten
Fifty-four plus six minus three equals fifty-seven

Write the name of the sign in each of these, p.30

$8 + 4$	plus or add
$16 ÷ 8$	divide or share
$3 × 6$	times or multiply
$44 ÷ 11$	divide or share
$18 − 6$	minus or subtract or take away
$17 + 22$	plus or add
$13 − 5$	minus or subtract or take away
$9 × 7$	times or multiply
$16 + 27$	plus or add
$36 − 20$	minus or subtract or take away
$14 ÷ 2$	divide or share
$16 − 7$	minus or subtract or take away
$2 + 19$	plus or add
$33 + 8$	plus or add

Write the sign + − × or ÷ next to each of these, p.30

fifteen divided by three	÷
eight plus twelve	+
24 divided by 6	÷
12 times 9	×
4 multiplied by 5	×
21 take away 13	−
102 added to 316	+
56 subtracted from 100	−
12 plus 8 plus 16	+
20 times 6	×
15 add 36	+
2 times 2 times 2	×
72 minus 24	−
18 divided by 2	÷
100 shared into 10 parts	÷

Add or multiply? p.31

$3 × 5 = 15$	$7 × 8 = 56$
$2 + 6 = 8$	$6 × 5 = 30$
$9 × 2 = 18$	$48 + 2 = 50$
$3 × 4 = 12$	$3 × 7 = 21$
$7 × 2 = 14$	$9 × 3 = 27$
$16 + 4 = 20$	$6 × 7 = 42$
$2 × 10 = 20$	$9 × 9 = 81$
$4 × 6 = 24$	$5 + 6 = 11$
$2 × 8 = 16$	$3 × 3 = 9$
$7 × 4 = 28$	

Put in the missing signs + −, p.31

$6 + 7 = 13$	$120 − 60 = 60$
$18 ÷ 2 = 9$	$24 ÷ 6 = 4$
$100 ÷ 10 = 10$	$12 × 3 = 36$
$17 − 5 = 12$	$16 ÷ 4 = 4$
$9 × 9 = 81$	$13 + 25 = 38$
$21 − 8 = 13$	$125 ÷ 25 = 5$
$4 × 5 = 20$	$6 + 12 + 3 = 21$
$158 − 16 = 142$	$10 + 5 − 3 = 12$
$15 × 4 = 60$	$2 × 2 × 2 = 8$

Which signs? p.32

1. The number of days in two weeks: $2 × 7$ or $7 + 7$

2. Jim is 9 today. How old will he be in 8 years time? $9 + 8$

3. My newspaper costs 45p a day. How much is this for 5 days? $5 × 45$ or $45 × 5$

4. We are walking on a 23-mile hike to Wike. The signpost says 'WIKE 16 miles'. How far have we travelled? $23 − 16$

5. A block of flats has 14 floors. There are 84 flats altogether. How many on each floor? $84 ÷ 14$

6. David's uncle is 41. He is 22 years older than David. How old is David? $41 − 22$

7. Aisha has 24 CDs in her collection. Last week she lost 5 of them. How many has she left? $24 − 5$

8. There are 6 squares in a bar of chocolate. How many squares in 8 bars? $6 × 8$

Write the multiplication sums that match the addition sums on the left hand side, p.36

ADD +	MULTIPLY ×
6 + 6	6 × 2 or 2 × 6
4 + 4 4 4	4 × 4
7 + 7 7	3 × 7 or 7 × 3
5 + 5 5	5 × 3 or 3 × 5
10 + 10 10 10	4 × 10 or 10 × 4

Breaking up numbers in different ways, p.37

4 lots of 2 4 x 2	2 lots of 4 2 x 4
5 lots of 3 5 x 3	3 lots of 5 3 x 5
4 lots of 5 4 x 5	5 lots of 4 5 x 4

Factors, p.38

48 = 4 × 12 = 2 × 2 × 2 × 2 × 3

33 = 3 × 11

100 = 10 × 10 = 2 × 5 × 2 × 5

24 = 3 × 8 or 2 × 12 or 4 × 6 = 2 × 2 × 2 × 3

16 = 4 × 4 or 2 × 8 = 2 × 2 × 2 × 2

39 = 3 × 13

Multiplying is quicker than adding and shorter to write, p.40

5 + 5 + 5 + 5 = 20	4 × 5 = 20
7 + 7 + 7 + 7 + 7 = 35	5 × 7 = 35
8 + 8 + 8 = 24	3 × 8 = 24
9 + 9 + 9 + 9 + 9 + 9 + 9 = 63	7 × 9 = 63
2 + 2 + 2 + 2 + 2 + 2 = 12	6 × 2 = 12
6 + 6 = 12	2 × 6 = 12
4 + 4 + 4 + 4 = 16	4 × 4 = 16
12 + 12 + 12 + 12 + 12 + 12 = 72	6 × 12 = 72
three sixes are eighteen	3 × 6 = 18
five fours are twenty	5 × 4 = 20
nine eights are seventy-two	9 × 8 = 72
fifteen times ten makes a hundred and fifty	15 × 10 = 150

Notes

Contents

Introduction

Addition, subtraction, multiplication and division are called the four rules or sometimes the four operations in arithmetic.

They underpin a lot of mathematics so if you can add, subtract, multiply and divide using whole numbers, decimals and fractions, and if you understand place value, you will be well on your way to becoming at ease with the maths in your life.

This book focuses on addition, subtraction, multiplication and division with whole numbers. Addition, subtraction, multiplication and division with decimals and fractions are in Book 3. Place value is covered in Books 1 and 3.

How well can you add?

There are different ways of saying and writing addition and subtraction calculations.
They can either be written along a line or with numbers above one another. The sign can
be written either on the right or the left of the numbers – you can choose:

3 + 4 =	6 + 6 =	8 + 7 =	5 + 9 =

3+	9+	7+	8+
4	8	6	8
7	2	3	3
	1	4	2

14+	26+	54+	36+
5	33	20	43

17+	64+	53+	77+
29	58	78	65

315+	452+	506+	375+
137	249	84	460

128+	607+	529+	658+
251	320	18	99
436	392	753	776

532+	7520+	3479+	2496+
1807	165	1009	4032
321	3178	3795	1006
14	4221	4556	28

If you can do this: **3+** and this: **61+**
                4   **25**

how about this?   **28+**
                   34

See the section on place value in Book 1, which explains how numbers are organised.

Start from the right hand side as usual:

Say **8 + 4 = 12**

Put down the **2** under the U (units) column.

Carry the **1** (that means add it to the next column to the left, the tens or T column).

$$\begin{array}{cc} T & U \\ 2 & 8 \\ 3 & 4 \\ \hline & 2 \\ 1 \end{array} +$$

Now add the numbers on the left.

Say **2 + 3 = 5**

and **1 + 5 = 6**

You add on the number you carried.

Put down the **6** under the tens column.

The answer is **62**.

$$\begin{array}{cc} T & U \\ 2 & 8 \\ 3 & 4 \\ \hline 6 & 2 \\ 1 \end{array} +$$

Now try these:

59+	**37+**	**88+**	**65+**
42	**53**	**17**	**28**

Adding single numbers

Add these numbers together.

Circle round the sums that have the same answer. What do you notice about them?

5 + 5 =	7 + 3 =	1 + 9 =	2 + 8 =	3 + 7 =
2 + 2 =	4 + 6 =	6 + 4 =	9 + 1 =	8 + 2 =
8 + 1 =	2 + 5 =	3 + 6 =	5 + 3 =	7 + 1 =
7 + 2 =	4 + 4 =	5 + 4 =	7 + 7 =	9 + 8 =
9 + 6 =	5 + 8 =	6 + 7 =	8 + 6 =	3 + 3 =
9 + 9 =	6 + 6 =	4 + 9 =	7 + 8 =	7 + 5 =

6+ 7 3	9+ 6 5	9+ 7 7	5+ 4 8	7+ 5 6
4+ 3 8	1+ 9 7	7+ 2 8	8+ 5 6	3+ 9 9
5+ 9 7	8+ 9 6	2+ 7 8	7+ 5 3	9+ 4 5
4+ 7 6 5	9+ 3 2 7	8+ 6 4 9	3+ 8 7 5	9+ 3 9 4

N1/E1.4, N1/E2.3

Adding double figures

T U	T U	T U	T U	T U
48+	66+	28+	72+	56+
13	15	14	19	26

69+	43+	69+	57+	88+
22	27	14	27	4

76+	35+	27+	39+	58+
18	49	47	32	18

N1/E2.3

Adding-up problems

1. My fare to work is £2.45p. How much does it cost me to go to work and come home again?

2. If this Tuesday is the 8th of May, what is the date next Tuesday?

3. My friend's rent has gone up by £20.70 a week. She used to pay £105. How much is it now?

4. The score at half-time was Arsenal 2, West Ham 0. West Ham scored 3 goals in the second half and Arsenal did not score again, so what was the score at the end of the match?

5. Morgan is 25, Jean is 6 years older, how old is Jean?

6. In the canteen I had a cheese sandwich for £1.82p and a cup of tea, 55p, for lunch. How much did my lunch cost altogether?

7. Fred's holiday was going to be from the 3rd June to the 17th June. Now he finds he's got to change the dates to begin and end 10 days later. What are the new dates?

8. Mia's three children each wanted something to eat. Kim wanted an ice cream costing 99p, Lee wanted a drink costing £1.18p and Ling wanted crisps costing 48p. How much did it all come to?

9. What year will it be in 5 years' time?

10. What is the date one week from today?

Adding three-figure numbers

HTU
264+
513
___

HTU
382+
116
___

HTU
523+
371
___

483+
513
___

276+
413
___

278+
321
___

287+
512
___

463+
234
___

862+
117
___

354+
625
___

642+
123
___

572+
315
___

521+
337
___

518+
261
___

327+
162
___

634+
312
___

776+
121
___

343+
514
___

How well can you take away?

There are different ways of saying and writing calculations. They can be written along a line or with numbers above one another and the sign on either the right or left.

$2 - 1 =$ $6 - 4 =$ $10 - 5 =$ $8 - 2 =$ $11 - 3 =$ $15 - 6 =$

$$
\begin{array}{r} 60- \\ 32 \\ \hline \end{array}
\qquad
\begin{array}{r} 95- \\ 87 \\ \hline \end{array}
\qquad
\begin{array}{r} 99- \\ 7 \\ \hline \end{array}
\qquad
\begin{array}{r} 16- \\ 13 \\ \hline \end{array}
$$

$$
\begin{array}{r} 80- \\ 78 \\ \hline \end{array}
\qquad
\begin{array}{r} 100- \\ 99 \\ \hline \end{array}
\qquad
\begin{array}{r} 250- \\ 125 \\ \hline \end{array}
\qquad
\begin{array}{r} 327- \\ 129 \\ \hline \end{array}
$$

$$
\begin{array}{r} 472- \\ 365 \\ \hline \end{array}
\qquad
\begin{array}{r} 588- \\ 298 \\ \hline \end{array}
\qquad
\begin{array}{r} 536- \\ 156 \\ \hline \end{array}
\qquad
\begin{array}{r} 371- \\ 203 \\ \hline \end{array}
$$

$$
\begin{array}{r} 318- \\ 89 \\ \hline \end{array}
\qquad
\begin{array}{r} 407- \\ 316 \\ \hline \end{array}
\qquad
\begin{array}{r} 894- \\ 776 \\ \hline \end{array}
\qquad
\begin{array}{r} 243- \\ 75 \\ \hline \end{array}
$$

$$
\begin{array}{r} 3604- \\ 1759 \\ \hline \end{array}
\qquad
\begin{array}{r} 2060- \\ 399 \\ \hline \end{array}
\qquad
\begin{array}{r} 58762- \\ 6898 \\ \hline \end{array}
\qquad
\begin{array}{r} 27908- \\ 8799 \\ \hline \end{array}
$$

$$15 - 3$$

$$14 - 2$$

$$13 - 3$$

$$19 - 8$$

$$27 - 13$$

$$25 - 14$$

$$28 - 11$$

$$29 - 18$$

$$36 - 21$$

$$37 - 12$$

$$35 - 21$$

$$34 - 22$$

$$58 - 31$$

$$54 - 41$$

$$56 - 24$$

$$55 - 22$$

$$64 - 51$$

$$62 - 41$$

$$65 - 42$$

$$67 - 16$$

$$94 - 42$$

$$88 - 67$$

$$78 - 67$$

N1/E2.3

Taking away – different methods

There are different ways of taking away when the top number is larger than the bottom number in one or more of the columns. These include taking away with borrowing, and using the counting on and decomposition methods.

When taking away, you need to know which number you are taking away from which.

This is unlike adding or multiplying, where the order of the numbers doesn't matter.

How to take away – borrowing method

$$\begin{array}{cc} T & U \\ 4\,^13 & - \\ {}^12\,6 & \\ \hline 1\,7 & \end{array}$$

This is one way of doing take-away sums.

Start from the right hand side as usual.

Say '**3** take away **6**, you can't do it'.

So borrow from the tens column. You need to borrow one lot of ten.

Put the **1** next to the **3** to make **13**.

Now say **13 − 6 = 7**.

Write the **7** down under the units column.

$$\begin{array}{cc} T & U \\ 4\,^13 & - \\ 2\,6 & \\ \hline 7 & \end{array}$$

You need to pay back the ten you borrowed.

Remember that, back in its own column, the ten is only worth one, so put **1** next to the **2** to make **3**, or cross out the **2** and write **3**.

$$\begin{array}{cc} T & U \\ 4\,^13 & - \\ {}^12\,6 & \\ \hline 7 & \end{array}$$

or

Now say **4 − 3 = 1**.

Write the **1** down under the tens column.

You've done it. The answer is **17**.

$$\begin{array}{cc} T & U \\ 4\,^13 & - \\ {}^3\cancel{2}\,6 & \\ \hline 1\,7 & \end{array}$$

Now try these:

$$\begin{array}{r} 62- \\ 37 \\ \hline \end{array} \qquad \begin{array}{r} 41- \\ 18 \\ \hline \end{array} \qquad \begin{array}{r} 93- \\ 25 \\ \hline \end{array} \qquad \begin{array}{r} 84- \\ 26 \\ \hline \end{array}$$

This method builds up numbers from the smallest to the largest to make round tens and hundreds.

How much do we need to add to **22** to make it up to **46**?

$$46-$$
$$22$$
———

Add **8** to make **22** up to **30**
add **16** to make **30** up to **46**

So the answer is **24**

$$8+$$
$$16$$
———
$$24$$

Do these in a similar way

$$51-$$
$$36$$
———

Add **4** to make **36** up to **40**
Add **11** to make **40** up to **51**

So the answer is **15**

$$4+$$
$$11$$
$$15$$
———

$$229-$$
$$115$$
———

Add **5** to **15** to make **20**
Add **80** to **120** to make **200**
Add **29** to make **200** up to **229**

So the answer is **114**

$$5+$$
$$80$$
$$29$$
———
$$114$$

$$342-$$
$$137$$
———

Add **3** to **37** to make **40**
Add **60** to **140** to make **200**
Add **142** to **200** to make **342**

So the answer is **205**

$$3+$$
$$60$$
$$142$$
———
$$205$$

Now try these:

$$84-$$
$$76$$
———

$$253-$$
$$166$$
———

$$405-$$
$$291$$
———

$$1562-$$
$$944$$
———

How to take away – decomposition method

This method splits the numbers into hundreds, tens and units.

No borrowing

$$\begin{array}{r} 37\, - \\ 14 \\ \hline 23 \end{array}$$

This can be written:

$$\begin{array}{r} 30 + 7\, - \\ 10 + 4 \\ \hline 20 + 3 \end{array}$$

Each part of the sum is worked on its own: $30 - 10 = 20$ $7 - 4 = 3$ $20 + 3 = 23$

Borrowing

$$\begin{array}{r} 37\, - \\ 18 \\ \hline \end{array}$$

This becomes:

$$\begin{array}{r} 30 + 7\, - \\ 10 + 8 \\ \hline \end{array}$$

but we can't take 8 from 7, so we split 37 in a different way, into $20 + 17$.

So

$$\begin{array}{r} 30 + 7\, - \\ 10 + 8 \\ \hline \end{array}$$

becomes:

$$\begin{array}{r} 20 + 17\, - \\ 10 + 8 \\ \hline 10 + 9 \end{array}$$

In short form, we write:

$$\begin{array}{r} {}^2\!\cancel{3}{}^1 7\, - \\ 1\ 8 \\ \hline 1\ 9 \end{array}$$

Follow these examples:

$$\begin{array}{r} 54\, - \\ 26 \\ \hline \end{array} \longrightarrow \begin{array}{r} 50 + 4\, - \\ 20 + 6 \\ \hline \end{array} \longrightarrow \begin{array}{r} 40 + 14\, - \\ 20 + \ 6 \\ \hline 20 + \ 8 \end{array} \longrightarrow \begin{array}{r} {}^4\!\cancel{5}{}^1 4\, - \\ 2\ 6 \\ \hline 2\ 8 \end{array}$$

$$\begin{array}{r} 372\, - \\ 159 \\ \hline \end{array} \longrightarrow \begin{array}{r} 300 + 70 + 2\, - \\ 100 + 50 + 9 \\ \hline \end{array} \longrightarrow \begin{array}{r} 300 + 60 + 12\, - \\ 100 + 50 + \ 9 \\ \hline 200 + 10 + \ 3 \end{array} \longrightarrow \begin{array}{r} 3\,{}^6\!\cancel{7}{}^1 2\, - \\ 1\ 5\ 9 \\ \hline 2\ 1\ 3 \end{array}$$

Now try these:

$$\begin{array}{r} 85\, - \\ 48 \\ \hline \end{array} \qquad \begin{array}{r} 42\, - \\ 17 \\ \hline \end{array} \qquad \begin{array}{r} 254\, - \\ 149 \\ \hline \end{array} \qquad \begin{array}{r} 526\, - \\ 218 \\ \hline \end{array} \qquad \begin{array}{r} 373\, - \\ 127 \\ \hline \end{array}$$

Borrowing when there are no tens

$$307- \quad \longrightarrow \quad 300 + 0 + 7- \quad \longrightarrow \quad 200 + 90 + 17- \quad \longrightarrow \quad {}^2 3 {}^9 0 {}^1 7-$$

$$\underline{\quad 49} \qquad \qquad \underline{\quad 40 + 9} \qquad \qquad \underline{\quad 40 + 9} \qquad \qquad \underline{\quad 4\ 9}$$

$$2\ 5\ 8$$

At stage 3 we have to borrow from the hundreds because there are no tens.
We take ten from the hundred we have borrowed to go with the 7, leaving 90.

So **307 = 300 + 7 = 200 + 100 + 7 = 200 + 90 + 17**

Try these:

$$\begin{array}{r} 204- \\ \underline{158} \end{array} \qquad\qquad \begin{array}{r} 508- \\ \underline{99} \end{array} \qquad\qquad \begin{array}{r} 301- \\ \underline{253} \end{array}$$

$$\begin{array}{r} 205- \\ \underline{88} \end{array} \qquad\qquad \begin{array}{r} 502- \\ \underline{425} \end{array}$$

N1/E3.2

Take away

Try these, using your preferred method of taking away.

61− 35	43− 28	72− 49	55− 47	37− 19

24− 16	81− 38	62− 17	45− 38	172− 89

423− 76	833− 189	602− 363	510− 297	304− 148

N1/E2.3, N1/E3.2

Taking away problems

1. Rafiq is 46. How old was he 14 years ago?

2. I left home with £20.50 in my purse. I spent £13.00 on groceries and 48p on a newspaper. How much did I have left?

3. The time in New York is 5 hours behind the time in Britain. When it's 11 o'clock in the morning in Britain, what time is it in New York?

4. What time is it now? What time was it a quarter of an hour ago?

5. My Great Grandmother died in 1986. She was born in 1918. How old was she when she died?

6. This month is ....

   5 months ago it was ....

7. Danny is 50, his sister is 9 years younger, how old is she?

8. In a room are 25 adults; 6 are men, how many are women?

9. Jamal got a pay rise of £16.00 a week which brought his pay up to £245.00. What did he get before the rise?

10. The average annual salary in the UK before tax is £20,919. Leroy's pay before tax is £35,000 per year. By how much is his pay above the national average?

$4 \times 5 =$ $2 \times 6 =$ $7 \times 8 =$ $5 \times 9 =$ $6 \times 6 =$ $12 \times 2 =$

18 × 3	26 × 4	33 × 5	69 × 6	20 × 5

84 × 7	62 × 7	49 × 6	98 × 8	88 × 6

104 × 4	326 × 7	230 × 6	345 × 9	28 × 10

35 × 30	271 × 11	198 × 11	307 × 12	450 × 12

586 × 15	349 × 21	727 × 34	89 × 64	608 × 60

337 × 48	750 × 29	1834 × 56	2072 × 37	158 × 99

326 × 150	872 × 206	1760 × 367	8121 × 286

Multiplication square

Use this multiplication square to check the times tables. For example, to find 4×5 go along the 4 line and down the 5 line, where the lines meet is where the answer is, in this case, $4 \times 5 = 20$.

You can also go along the 5 line and down the 4 line because $5 \times 4 = 20$ as well. The order does not matter when you multiply.

Look for similar examples in the multiplication square.

1	2	3	4	5	6	7	8	9	10	11	12
2	4	6	8	10	12	14	16	18	20	22	24
3	6	9	12	15	18	21	24	27	30	33	36
4	8	12	16	20	24	28	32	36	40	44	48
5	10	15	20	25	30	35	40	45	50	55	60
6	12	18	24	30	36	42	48	54	60	66	72
7	14	21	28	35	42	49	56	63	70	77	84
8	16	24	32	40	48	56	64	72	80	88	96
9	18	27	36	45	54	63	72	81	90	99	108
10	20	30	40	50	60	70	80	90	100	110	120
11	22	33	44	55	66	77	88	99	110	121	132
12	24	36	48	60	72	84	96	108	120	132	144

You can fill in your own multiplication square using the number square in Book 1.

This multiplication square goes up to 12 but you can make a multiplication square up to any number.

Use the multiplication square (if you need it) to fill in the blanks.

Circle round the pairs of sums that have the same answer. What do you notice about them?

$5 \times 9 =$ $3 \times 7 =$ $9 \times 10 =$ $6 \times 5 =$ $7 \times 6 =$

$8 \times 9 =$ $6 \times 8 =$ $4 \times 9 =$ $9 \times 8 =$ $9 \times 7 =$

$6 \times 9 =$ $8 \times 8 =$ $8 \times 7 =$ $7 \times 8 =$ $9 \times 5 =$

$4 \times 3 =$ $5 \times 4 =$ $9 \times 6 =$

Fill in the blanks.

$8 \times \underline{} = 64$ $7 \times \underline{} = 56$ $\underline{} \times 4 = 36$

$\underline{} \times 9 = 81$ $\underline{} \times 9 = 45$ $5 \times \underline{} = 35$

$\underline{} \times 9 = 63$ $7 \times \underline{} = 49$ $\underline{} \times 8 = 40$

Find pairs of numbers to fill in the blanks:

$\underline{} \times \underline{} = 60$ $\underline{} \times \underline{} = 21$ $\underline{} \times \underline{} = 18$

$\underline{} \times \underline{} = 24$ $\underline{} \times \underline{} = 16$ $\underline{} \times \underline{} = 18$

$\underline{} \times \underline{} = 24$ $\underline{} \times \underline{} = 16$ $\underline{} \times \underline{} = 33$

$\underline{} \times \underline{} = 35$ $\underline{} \times \underline{} = 54$ $\underline{} \times \underline{} = 48$

Multiplication is repeated addition

As we saw in Book 1, multiplying is a quick way of adding the same numbers together a number of times.

For example, $3 \times 5 = 15$

is the same as $5 + 5 + 5 = 15$

and $3 + 3 + 3 + 3 + 3 = 15$

which is the same as $5 \times 3 = 15$.

Whichever way round you do it, you should still get the same answer, so you can use this for checking.

Times tables

Fill in the answers on the following pages using the multiplication square if you need it.

You could also use repeated addition. For example 4×3 is the same as $3 + 3 + 3 + 3$. You can use this to check your times tables if you're not sure of them. If you're stuck you can always add on another of the numbers you're multiplying by.

Look for patterns in the times tables, for example, odd and even numbers, numbers ending in 5 or 0, etc.

The numbers that multiply together to give another number are called factors.
For example, in $2 \times 4 = 8$, 2 and 4 are factors of 8.

The answer you get when multiplying numbers together is called the product of those numbers, so 8 is the product of 2×4.

One times table:

$1 \times 1 = 1$ one times one is one

$2 \times 1 =$

$3 \times 1 =$

$4 \times 1 =$

$5 \times 1 =$

$6 \times 1 =$

$7 \times 1 =$

$8 \times 1 =$

$9 \times 1 =$

$10 \times 1 =$

$11 \times 1 =$

$12 \times 1 =$

Two times table:

$1 \times 2 = 2$ one times two is two

$2 \times 2 =$

$3 \times 2 =$

$4 \times 2 =$

$5 \times 2 =$

$6 \times 2 =$

$7 \times 2 =$

$8 \times 2 =$

$9 \times 2 =$

$10 \times 2 =$

$11 \times 2 =$

$12 \times 2 =$

Three times table:

$1 \times 3 = 3$ one times three is three

$2 \times 3 =$

$3 \times 3 =$

$4 \times 3 =$

$5 \times 3 =$

$6 \times 3 =$

$7 \times 3 =$

$8 \times 3 =$

$9 \times 3 =$

$10 \times 3 =$

$11 \times 3 =$

$12 \times 3 =$

Four times table:

$1 \times 4 = 4$ one times four is four

$2 \times 4 =$

$3 \times 4 =$

$4 \times 4 =$

$5 \times 4 =$

$6 \times 4 =$

$7 \times 4 =$

$8 \times 4 =$

$9 \times 4 =$

$10 \times 4 =$

$11 \times 4 =$

$12 \times 4 =$

N1/E2.5, N1/E3.4, N1/E3.5

Five times table:

$1 \times 5 = 5$ one times five is five
$2 \times 5 =$
$3 \times 5 =$
$4 \times 5 =$
$5 \times 5 =$
$6 \times 5 =$
$7 \times 5 =$
$8 \times 5 =$
$9 \times 5 =$
$10 \times 5 =$
$11 \times 5 =$
$12 \times 5 =$

Six times table:

$1 \times 6 = 6$ one times six is six
$2 \times 6 =$
$3 \times 6 =$
$4 \times 6 =$
$5 \times 6 =$
$6 \times 6 =$
$7 \times 6 =$
$8 \times 6 =$
$9 \times 6 =$
$10 \times 6 =$
$11 \times 6 =$
$12 \times 6 =$

Seven times table:

$1 \times 7 = 7$ one times seven is seven
$2 \times 7 =$
$3 \times 7 =$
$4 \times 7 =$
$5 \times 7 =$
$6 \times 7 =$
$7 \times 7 =$
$8 \times 7 =$
$9 \times 7 =$
$10 \times 7 =$
$11 \times 7 =$
$12 \times 7 =$

Eight times table:

$1 \times 8 = 8$ one times eight is eight
$2 \times 8 =$
$3 \times 8 =$
$4 \times 8 =$
$5 \times 8 =$
$6 \times 8 =$
$7 \times 8 =$
$8 \times 8 =$
$9 \times 8 =$
$10 \times 8 =$
$11 \times 8 =$
$12 \times 8 =$

Nine times table:

$1 \times 9 = 9$ one times nine is nine
$2 \times 9 =$
$3 \times 9 =$
$4 \times 9 =$
$5 \times 9 =$
$6 \times 9 =$
$7 \times 9 =$
$8 \times 9 =$
$9 \times 9 =$
$10 \times 9 =$
$11 \times 9 =$
$12 \times 9 =$

Ten times table:

$1 \times 10 = 10$ one times ten is ten
$2 \times 10 =$
$3 \times 10 =$
$4 \times 10 =$
$5 \times 10 =$
$6 \times 10 =$
$7 \times 10 =$
$8 \times 10 =$
$9 \times 10 =$
$10 \times 10 =$
$11 \times 10 =$
$12 \times 10 =$

Eleven times table:

$1 \times 11 = 11$ one times eleven is eleven
$2 \times 11 =$
$3 \times 11 =$
$4 \times 11 =$
$5 \times 11 =$
$6 \times 11 =$
$7 \times 11 =$
$8 \times 11 =$
$9 \times 11 =$
$10 \times 11 =$
$11 \times 11 =$
$12 \times 11 =$

Twelve times table:

$1 \times 12 = 12$ one times twelve is twelve
$2 \times 12 =$
$3 \times 12 =$
$4 \times 12 =$
$5 \times 12 =$
$6 \times 12 =$
$7 \times 12 =$
$8 \times 12 =$
$9 \times 12 =$
$10 \times 12 =$
$11 \times 12 =$
$12 \times 12 =$

N1/E2.5, N1/3.4, N1/E3.5

One times table:

Multiplying by 1 does not change the number being multiplied, for example:
$6 \times 1 = 6$

Two times table:

Multiplying by **2** is doubling a number.

Three times table:

Note that if you add together the figures after the equals signs, they are also multiples of 3.
For example, $7 \times 3 = 21$; $2 + 1 = 3$.
You can use this to see if any number is a multiple of **3**.

Four times table:

You can multiply by **4** by doubling and then doubling again. For example, 3×4 is the same as 3 doubled, which is **6**, and then
6 doubled, which is **12**, so $3 \times 4 = 12$.

Five times table:

Multiples of **5** always end in **5** or **0**.

Six times table:

To multiply by **6** multiply by **3** and then double your answer.

Eight times table:

You can multiply by **8** by doubling, doubling again and then doubling again. For example, 3×8 is the same as **3** doubled, which is **6**, and then **6** doubled, which is **12**, and then **12** doubled, which is **24**, so $3 \times 8 = 24$.

Alternatively, if you know your four times table you can double from that, in which case you only have to double once. For example,
3×8 is the same as 3×4 doubled;
3×4 is **12**, **12** doubled is **24**.

Nine times table:

You can add on **10** and take away **1** to find multiples of **9**. You can also use your fingers. Think of your fingers as labelled **1**, **2**, **3**, etc. starting from the thumb of your left hand. If you want **4** $\times$ **9**, bend your 4th finger down. The number of fingers to the left tells you how many tens in the answer and the number of fingers to the right tells you the number of ones. In this case you'll have **3** tens and six ones: **36**.

Notice the patterns in the nine times table. For example, the digits in the multiples of **1** $\times$ **9** and so on, up to **10** $\times$ **9**, always add up to **9** (try it and see). Also, as one digit goes up one, the other digit goes down one, so **2** $\times$ **9** = **18** and **3** $\times$ **9** = **27** and so on.

Ten times table:

An easy way to multiply by **10** is to put a **0** on the end of the number you started with, for example, **8** $\times$ **10** = **80** (see the section on 'Multiplying by 10' in this book and the sections on 'Place Value' in Book 1 to find out why this works). This means any number ending in **0** is a multiple of **10**.

Eleven times table:

Notice the patterns in the eleven times table.

Twelve times table:

12 is called a dozen. **12** times **12** is **144**, or a gross.

Times table test sheet

Fill in the gaps:

$2 \times 4 = 8$

$3 \times 2 =$

$7 \times 3 =$

$2 \times \quad = 10$

$3 \times \quad = 30$

$5 \times 11 =$

$3 \times 3 =$

$4 \times \quad = 16$

$5 \times \quad = 20$

$\quad \times 5 = 30$

$\quad \times 4 = 32$

$7 \times \quad = 42$

$4 \times \quad = 28$

$6 \times 6 =$

$5 \times \quad = 25$

$7 \times 8 =$

$4 \times 10 =$

$6 \times 3 =$

$2 \times \quad = 18$

$7 \times \quad = 35$

$7 \times 12 =$

$10 \times \quad = 50$

$9 \times 9 =$

$\quad \times 8 = 40$

$9 \times \quad = 54$

$6 \times 8 =$

$8 \times \quad = 16$

$8 \times 7 =$

$5 \times 9 =$

$\quad \times 7 = 70$

$8 \times 9 =$

$7 \times 7 =$

$4 \times \quad = 36$

$7 \times \quad = 63$

$10 \times \quad = 90$

$11 \times \quad = 110$

$1 \times \quad = 10$

$4 \times 8 =$

$6 \times 9 =$

$2 \times \quad = 12$

$\quad \times 3 = 27$

$8 \times \quad = 64$

$10 \times \quad = 60$

$8 \times 10 =$

$4 \times 12 =$

Square numbers and square roots

Square numbers

When you multiply a number by itself, the answer is called a square number.

For example, **16** is a square number because $4 \times 4 = 16$

Statements with square numbers can be written like this: $4 \times 4 = 16$
or like this: $4^2 = 16$

The small ^{2} next to the **4** means the **4** is squared, or multiplied by itself (see the section on 'Powers and indices' in this book for more on this).

Square roots

You can also say that **4** is the square root of **16**

In other words, it is the number which, when multiplied by itself, makes **16**

Square root is written $\sqrt{}$, so you could write $\sqrt{16} = 4$

Look for other square numbers on the multiplication square.

What do you notice about them?

Write down all the square numbers you can think of, setting them out to show the numbers that multiply together to make them.

For example: $5 \times 5 = 25$ $5^2 = 25$ and so on.

Then, using the same square numbers, set them out to show the square roots.

For example, $\sqrt{25} = 5$

Short multiplication

$42 \times$	$53 \times$	$67 \times$	$35 \times$	$84 \times$
6	4	5	7	9

$63 \times$	$21 \times$	$70 \times$	$97 \times$	$123 \times$
6	8	3	6	6

$518 \times$	$704 \times$	$806 \times$	$902 \times$	$510 \times$
9	7	5	7	8

N1/E3.4

Short multiplication problems

1. 6 people went to the cinema. Their seats cost £5.50 each. How much did they pay altogether?

2. How much do 10 first class stamps cost?

3. My return fare to work costs me £3.75p. How much do I pay for 5 days travel?

4. I want 6 shrubs for my garden. Each one costs £5.99. How much must I pay altogether?

5. How many flats are there in a block with 13 floors and 2 flats on each floor?

6. There are 3 months to go before my holiday. If I save £25 a month how much will I have by the time I go?

7. I need 25 tiles to tile my bathroom floor. Each tile costs £1.60p. How much will it cost to do the job?

8. 1 litre of petrol costs 80p. How much will 6 litres come to?

9. A round of drinks at the pub cost £8.70. That was for 3 bottles of beer at £2.35p. Was the price charged correct? If not, what should it have been?

The rule is 'put **0** on the end' – but why?

$$9 \times 10 = 90 \qquad 7 \times 10 = 70$$
$$25 \times 10 = 250 \qquad 181 \times 10 = 1810$$

The number **25** stands for **2** tens and **5** units and is written in columns like this:

Remember, each column to the left has the value of **10** times the previous one.

When we multiply by **10**, we want each figure to stand for something **10** times bigger.

Th	H	T	U
		2	5

$$10 = 10 \times 1$$
$$1000 = 100 \times 10$$
$$100 = 10 \times 10$$

So, hundreds become thousands
    tens become hundreds
    units become tens.

That is, the figures move **1** column to the left.

It looks like this:

original number

'10 times bigger'

The units column is now empty.

We must show this by putting in **0** to hold the place and show the difference between **25** and **250**.

$$350 \times 10 = 3,500 \qquad 3,785 \times 10 = 37,850$$

$$1,972 \times 10 = 19,720 \qquad 118 \times 10 = 1,180$$

Th	H	T	U
		2	5
	2	5	
	2	5	0

N1/E3.1, N1/L1.4

15 × 10	15 × 20	15 × 30	15 × 40	15 × 50	15 × 60
23 × 10	23 × 20	23 × 30	23 × 40	23 × 50	23 × 60
31 × 10	31 × 30	31 × 50	31 × 70	31 × 80	31 × 90
42 × 20	42 × 40	42 × 50	42 × 60	42 × 70	42 × 90
56 × 10	56 × 30	56 × 40	56 × 60	56 × 70	56 × 80
67 × 10	67 × 20	67 × 30	67 × 40	67 × 50	67 × 60

$$32 \times 100 = 3200$$

32 stands for **3** tens and **2** units.

In columns, it looks like this:

To multiply **32** by **100** we must make each figure stand for something **100** times bigger.

So, units become hundreds
    tens become thousands
    hundreds become ten thousands.

That is, the figure moves two columns to the left.

This leaves two empty columns – the units and tens.

We show this by putting **0** in each to hold the place and show the difference between **32** and **3200**.

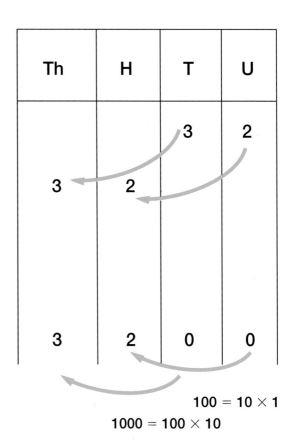

100 = 10 × 1
1000 = 100 × 10

So the rule is 'put two noughts on the end' because the noughts (or zeros) are 'place holders' – keeping the numbers in the right places.

$$6 \times 100 = 600$$

$$12 \times 100 = 1{,}200$$

$$47 \times 100 = 4{,}700$$

$$321 \times 100 = 32{,}100$$

$$5{,}386 \times 100 = 538{,}600$$

How to do long multiplication

Long multiplication can be used when you need to multiply by a large number.

It works by separating the tens, units etc., and doing the multiplication in stages, then adding the resulting numbers together. For example:

48 ×					48 ×	
21					21	
48 +	1	times +	or		960	20 times +
960	20	times			48 +	1 times
1008	21	times			1008	21 times

You can multiply by the 1 first and then by 20, or the other way round.

Use 0 as a place holder and then multiply by the 2 in 21, because the 2 means 2 tens, or 20.

Now try these:

65 ×	35 ×	79 ×	58 ×	86 ×	44 ×
24	47	35	19	34	29

Long multiplication

66 ×	38 ×	79 ×	87 ×	94 ×
27	45	4	26	35

53 ×	29 ×	71 ×	68 ×	77 ×
17	46	32	51	36

89 ×	62 ×	304 ×	210 ×	781 ×
25	28	31	16	45

Long multiplication problems

1. A woman earns £90.00 a week in her part-time job, how much is that a year?

2. I worked out the area of the floor of my kitchen to be 140 square feet. It measures 10 feet by 12 feet, was I right?

3. At the Cash and Carry 50 tins of dog food cost £20.00. In my local shop the same brand costs 58p a tin, how much would I save by buying in bulk?

4. 20 cigarettes a day is how many in a year?

5. Alun bought 12 boxes of bedding plants with 15 plants in each box. How many plants is that altogether?

6. Netta earned £75.00 a day for 11 days as a temporary secretary, how much did she make altogether?

7. Mike bet Dave that there were more than 1000 holes in a piece of pegboard. There were 24 rows of 12 holes, who won the bet?

8. My rent is £140.00 a week, how much do I pay in a year?

9. I have seen a washing machine on sale for £499.00. I could buy it on hire purchase for 12 monthly payments of £44.00. How much would this come to?

10. My living room measures 4 metres by 5 metres. I want to buy some carpet for it and I've seen some I like at £18.95 a square metre. Approximately how much would it cost to carpet the room?

Strategies for multiplication with whole numbers

Look for patterns and work from what you know. There are lots of ways you can work things out in your head, or you may prefer to write things down or use a calculator.

You can use different methods to check your working out, for example, by using a calculator to check your written calculation, or to check your estimated answer.

To multiply by **2**, split the number into tens, units, etc., and double each figure, for example, **2 × 27** is double **20** plus double **7**, i.e. **40** + **14**, which makes **54**.

To multiply by **4**, double and double again, for example, if you want to know **4 × 20**, start by doubling **20**, which is **40**, then double again: **2 × 40** = **80**.

To multiply by **9**, start by multiplying by **10** and then subtract by the number of nines you need, like this…

Say you want to know **5 × 9**.
Multiply **5 × 10**, which gives you **50**, then take away **5**.

If you want to know **7 × 9**, multiply **7 × 10** (which equals 70) and take away **7**, and so on.

This method exploits the fact that nine is one less than ten and once you understand place value (see 'Place value' in Book 1) it's easy to multiply by 10.

To multiply by **10**, the rule is 'stick a **0** on the end'. To understand why this works, look at the sections on 'Place value' in Book 1 and the worksheet on page 27 of this book.

To multiply by **100** or **1000** or **10,000** the same principle applies. You put the same numbers of 0s on the right hand end of the number as there are 0s in **100**, **1000**, etc. For example, **5 × 100** = **500**; **23 × 10,000** = **230,000** and so on.

Factors, factorising and prime numbers

Factors and factorising

Factors are numbers which multiply together to make another number: $2 \times 3 = 6$.
In this case, 2 and 3 are factors of 6 because they multiply together to make 6.

Finding the factors of a number is called factorising.

Find the factors of these numbers (the multiplication square on p.16 is helpful):

18	**25**	**9**	**14**	**21**

Prime numbers

Prime numbers are numbers which cannot be broken down at all because they only divide by themselves and 1, for example:

2   **3**   **5**   **7**   **11**   **13**   **17**   **19** etc.

Write down the rest of the prime numbers up to 100.

Prime factors

Prime factors are the smallest numbers which a number can be broken down into. Prime factors are always prime numbers.

There may be several ways of breaking a number down into factors (factorising). When you can't break it down any further, you know you've found the prime numbers. This is called decomposition into prime factors. For example:

$$20 = 4 \times 5 \text{ and } 20 = 2 \times 10 \text{ and } 20 = 2 \times 2 \times 5$$

2 and 5 are prime factors of 20, because they are the smallest numbers which multiply together to make 20.

Break down these numbers into prime factors:

16   **12**   **18**   **21**   **30**   **27**   **24**   **25**   **36**

Factorising is a useful skill. It helps you to become familiar with numbers and how they work. It's used in many operations, for example, calculations with fractions (see Book 3).

Multiplying and dividing

Multiplying and dividing are 'two sides of the same coin'.

Look at the section on factorising in this book. **5** and **3** are factors of **15**, they multiply together to give **15**.

This means they also divide into **15** equally, so **15 ÷ 5 = 3** and **15 ÷ 3 = 5**.

Starting with 15, you can write down:

$$3 \times 5 = 15 \qquad 5 \times 3 = 15 \qquad 15 \div 5 = 3 \qquad 15 \div 3 = 5$$

Try that with these numbers using different pairs of factors:

| **12** | **14** | **18** | **24** | **56** | **35** | **60** | **33** |

Division is the opposite, or inverse, of multiplication, in other words, division undoes the multiplication process and multiplication undoes the division process.

N1/L2.2

Patterns in numbers: multiplying and dividing

Look for patterns in numbers and think about what they mean.

For example, think of a whole number ending in 0. The 0 in that position tells you that you must be able to divide it by 10.

Similarly, any number ending in 5 or 0 must be divisible by 5 (try it and see).

Any even number can be divided by 2 (see 'Odd and even numbers' in Book 1).

Multiplying and dividing by 3. If, when you add the digits in a number together (this is also called summing the digits), you can divide the total by 3 with no remainder, then the number you started with is also divisible by 3. Take 18 for example; add the 1 and the 8 **1 + 8 = 9**; divide the answer (9) by 3: **9 ÷ 3 = 3**, so you know that 18 is divisible by 3. Try it with other numbers.

Multiplying and dividing by 9. The same principle works for checking whether a number can be divided by 9. If, when you add the digits in a number together, you can divide the total by 9 with no remainder, then the number you started with is also divisible by 9. Take 36 for example; add the 3 and the 6: **3 + 6 = 9**; divide the answer (9) by 3: **9 ÷ 3 = 3**, so you know that 36 is divisible by 9. Try it with other numbers.

Powers and indices

Indices, or powers, tell you how many times to multiply a number by itself. Indices is the plural form of index.

Index numbers are written to the top right of the number to which they refer. For example,
3^2 means 3×3 3^3 means $3 \times 3 \times 3$, and so on.

You say three squared, or three to the power two, for 3^2, and three cubed, or three to the power three, for 3^3. For indices greater than 3 you say, for example, three to the power four (3^4), three to the power five (3^5), and so on.

Similarly, 10^2 is 100, 10^3 is 1000, 10^4 is 10,000, 10^5 is 100,000, 10^6 is 1,000,000, and so on. The number 10^{100}, or 10 multiplied by itself 100 times, is called a googol. A googol is written as 1 followed by 100 zeros. If you try to write 10^{100} out in full, you'll see how much more efficient it is to use indices.

It's important not to mix up, for example, 5^3 with 5×3.

$$5^3 = 5 \times 5 \times 5 = 125 \qquad \text{while} \quad 5 \times 3 = 15$$

Write these in full:

6^3 5^4 2^3 9^8 7^2 8^5 3^3 4^2 15^3 12^{10}

How well can you divide?

$10 \div 2 =$ $15 \div 3 =$ $16 \div 4 =$ $20 \div 5 =$ $66 \div 6 =$

$2 \overline{)26}$ $3 \overline{)636}$ $4 \overline{)408}$ $5 \overline{)550}$ $3 \overline{)999}$

$4 \overline{)250}$ $3 \overline{)810}$ $8 \overline{)656}$ $8 \overline{)306}$ $4 \overline{)7790}$

$20 \overline{)76}$ $23 \overline{)46}$ $18 \overline{)360}$ $12 \overline{)993}$ $12 \overline{)376}$

There are different ways of writing and saying division sums. You can say: six divided by two equals three, or two into six goes three times, or six over two is three.

$$6 \div 2 = 3$$

$$2\overline{)6}^{\,3}$$

$$\frac{6}{2} = 3 \quad \text{or} \quad 6/2 = 3$$

In the case of $\frac{6}{2}$ we say: six over two equals three. The line between 6 and 2 means divided by (the top number, 6, is divided by the bottom number, 2).

It's no coincidence that this looks like a fraction (see Book 3 for work on fractions). The ÷ sign is an abstract representation of a fraction, with the top number divided by the bottom number, as in a fraction.

Write these division sums out in the different ways:

$$12 \div 4 = 3$$

$$6\overline{)42}^{\,7}$$

$$\frac{18}{9} = 2$$

$$36 \div 9 = 4$$

Short division

$$12 \div 3 = 4$$

We say: twelve divided by three makes four.

Take **12** counters, divide them into **3** groups with the same number of counters in each.

Each group should have **4** counters.

This should remind you of what happens when you multiply, you would say:

$$3 \times 4 = 12$$

Three times four makes twelve, or three fours make twelve.

When you divide you start with the biggest number and divide it up into groups of the same size. Divide is another name for sharing.

Try these division sums (use the counters if you like).

$$15 \div 3 =$$

$$10 \div 2 =$$

$$9 \div 3 =$$

$$14 \div 2 =$$

$$20 \div 4 =$$

How to divide

$3 \overline{)516}$ **We say: 3** into **5** hundred and **16**

Start at the left-hand side.

Say: **3** into **5** goes **1** and **2** left over.

$$\begin{array}{r} 1 \\ 3\overline{)516} \end{array}$$

The question is, what do you do with the 2 left over?

You could write the sum out in the same way as it's done in the long division method worksheet in this book.

There is a quicker way though, which some people prefer.

This is how you do it:

Put the **2** with the **1** in the **516** like this:

$$\begin{array}{r} 1\ 7 \\ 3\overline{)5^216} \end{array}$$

Now say, **3** into **21** goes **7**, put the **7** above the **1**.

Now say, **3** into **6** goes **2**, put the **2** above the **6**.

$$\begin{array}{r} 1\ 72 \\ 3\overline{)5^216} \end{array}$$

The answer is **172**.

Dividing when there is a 0 in the number

You do this in exactly the same way:

$4 \overline{)408}$

You say: 4 into 4 goes 1 with nothing left over, put the 1 above the 4 in 408.

Then say: 4 into 0 goes 0, put the 0 above the 0 in 408.

$$\begin{array}{r} 102 \\ 4\overline{)408} \end{array}$$

Then: 4 into 8 goes 2, put the 2 above the 8 in 408.

It's very important to put the 0 in the answer, otherwise your answer will be much too small (12, when it should be 102).

You can check your answer by multiplying the answer by the number you were dividing by: in this case $102 \times 4 = 408$ so you can see the answer is correct.

What do you do if there is a remainder?

There are three different ways of writing a remainder in a division sum. You choose the most appropriate way for the sum you are doing.

For example: $5\overline{)28}$

The answer is 5 and 3 left over, because $5 \times 5 = 25$ and 28 is 3 more than 25.

You can write that like this:

$$5\overline{)28}^{\ 5\ r3}$$

Some people put R or rem. for remainder.

Alternatively, you can write it like this, as a fraction:

$$5\frac{3}{5}$$

Or you can turn the remainder into a decimal like this:

$$5\overline{)28.^30}^{\ 5.6}$$

Put a decimal point after the 28 and after the 5 in the answer. Put the 3 which is left over with the 0 in 28.0 and say 5 into 30 goes 6.

The answer is **5.6**.

If necessary, you could have gone on putting 0s after the decimal point until the answer came out exactly. Look at the 'Rounding off' page in Book 3 to see how to deal with the decimals.

If you use a calculator to do division sums you will find that it shows any remainder as decimals. If you find decimals confusing, look at the pages on decimals in Book 3.

Fill in the blanks

Use the multiplication square (if you need it).

$12 \div 3 =$ $24 \div 6 =$ $18 \div 9 =$ $27 \div 3 =$

$15 \div 5 =$ $18 \div 6 =$ $25 \div 5 =$ $36 \div 6 =$

$42 \div 7 =$ $63 \div 9 =$ $100 \div 10 =$ $88 \div 8 =$

N1/E3.6, N1/L1.3

Divide

$4\overline{)284}$ $5\overline{)450}$ $6\overline{)426}$ $7\overline{)560}$ $4\overline{)3916}$

$6\overline{)600}$ $9\overline{)819}$ $3\overline{)153}$ $5\overline{)205}$ $3\overline{)450}$

$2\overline{)184}$ $3\overline{)999}$ $2\overline{)246}$ $7\overline{)497}$ $7\overline{)6342}$

$2\overline{)868}$ $6\overline{)4812}$ $9\overline{)7218}$ $7\overline{)553}$ $5\overline{)732}$

N1/L1.3

Short division problems

1. Four of us went for a Chinese meal. The bill came to £80.00. How much did we each have to pay?

2. A 3-carton pack of orange juice costs £1.26 and 1 carton costs 45p. Is it cheaper to buy a 3-carton pack or 3 separate cartons?

3. Jagdish saves £5.00 a month. How long will it take him to save £145.00 for a camera?

4. Ming-Li wants to make some shelves to fit in an alcove 3 foot wide. How many shelves can she make out of a plank 8 foot long?

5. Bob needs to lose 10 kilograms. If he manages to lose 2 kilograms a week, how long will it take him to reach his target weight?

6. I brought 24 small bars of chocolate to share between 10 children at a party. How many can each child have? Will there be any left over?

7. A pop music programme on the radio lasts for half an hour. How many records lasting 3 minutes each can be played in that time?

8. George drove 100 miles in 5 days. How far did he go each day on average?

9. At the moment my rent is £95.00 a week, and I'm thinking of moving to a slightly bigger flat where the rent is £398.00 a month (4 weeks). Would I save money if I moved?

10. A pack of 6 bars of soap costs £2.40. How much is that for each bar?

How to do long division

Long division is dividing by numbers which are more than **10** or **12**, numbers we don't learn tables for.

For example: **575** divided by **25**

The number you divide by is called the divisor, in this case, **25**, and the number you are dividing is called the dividend, in this case the dividend is **575**.

Set it out like this: $25\overline{)575}$

We break this down into several smaller calculations.

Start from the figure at the beginning of the dividend.

Can we divide **25** into **5**? Answer: No.

Take the next figure, **7**, with the **5** to make **57**.

Can we divide **25** into **57**? Answer Yes.

How many times? Work this out in rough:

is it...? **1 × 25 = 25**
or?   **2 × 25 = 50**

It must be **2 × 25**, because this gives the answer nearest to **57** which is not too big. To find how many are left over, subtract **50** from **57**. So, **25** into **57** goes twice, with **7** left over.

The calculation so far:

$$
\begin{array}{r}
2 \\
25\overline{)575} \\
-50 \\
\hline
7
\end{array}
$$

Now we move on to the next figure, **5**, which joins the **7** to make **75**.

$$
\begin{array}{r}
2 \\
25\overline{)575} \\
-50 \\
\hline
75
\end{array}
$$

Can we divide **25** into **75**?
Answer: yes.

How many times?

Work this out as before.

1 × 25 = 25

2 × 25 = 50

3 × 25 = 75

We see that **3 × 25 = 75**.

So **25** goes into **75**, **3** times with nothing left over.

Set this out:

```
           23
     25) 575
        -50
         75
        -75
         00
```

There are no more figures left in the original number **575**, and the sum is finished. The final answer is **23**.

575 ÷ 25 = 23

or **575** divided by **25** equals **23**

or **25** goes into **575**, **23** times exactly.

Long division problems

1. There are 400 seats in the theatre in 25 rows. How many in each row?

2. I ordered 200 rolls for a wedding reception for 50 guests. I reckoned this allowed 3 rolls each. Was I right?

3. In 32 cases there are 768 apples. How many in each case?

4. I have 256 chocolates to pack into 15 boxes. How many in each box? Are there any left over?

5. I have to pack 72 light bulbs in boxes, 24 to each box. How many boxes will I need? Are there any left over?

6. I have to order 500 bottles of beer for a party. They come in crates of 36. How many crates do I need?

7. 500 people are going to a football match. How many coaches do they need to hire, if each coach seats 40 people?

8. I earn £14,680.00 a year. How much is this per week?

9. I want to save about £450.00 a year. Roughly how much do I need to save each month?

Estimation

You do not always need an exact answer. It's often more important to get a rough idea.

You do need to decide how accurate you should be, and that will depend on the context. For example, if you have plenty of money and a credit card, you don't need to check how much cash you've got before going shopping.

Estimation, also known as approximation, is a very useful skill. It's important to develop your number sense so that you can estimate the answer to a calculation and know whether it's sensible or not. This is especially important when you're using a calculator, as it's easy to make a mistake keying in numbers.

You can round numbers up or down to make them easier to deal with.

For example, **179** can be rounded up to **180**, and **£25.15** can be rounded down to **£25**. If you're dealing with large numbers, you may need to round up or down to the nearest hundred, thousand, million, etc.

The general rule is, if the number is at or beyond the halfway point, then you round up, if it's less than halfway, you round down. For example, **267** would be rounded up to **270** or **300**, depending on the degree of accuracy required; **133** would be rounded down to **130** or **100**, again, depending on the degree of accuracy required.

Again, it all depends on the context, if you're shopping with cash and don't want to spend more than the amount you've got with you, then you need to round up and not down to be sure you don't overspend.

In some trades, estimates are very important, for example, a builder will give a prospective customer an estimate of the cost of work to be done. The actual cost may turn out to be higher or lower than the estimate.

To find the **average**, or mean, of a group of numbers, add together all the numbers, then divide by however many numbers there are.

For example, find the average of **12, 5, 18, 14, 26**.

First add all the numbers:

$$12+$$
$$5$$
$$18$$
$$14$$
$$\underline{26}$$
$$\underline{75}$$

Then divide the total by **5** because there are **5** numbers.

$$5\overline{)75}\;\;\frac{15}{}$$

The answer to this sum is the average of all the numbers, in this case, **15**.

So the average of **12, 5, 18, 14,** and **26** is **15**.

The average may not turn out to be a whole number, as it has done here.

The average is a number that typifies a group of numbers, levelling them out. This diagram shows how it works:

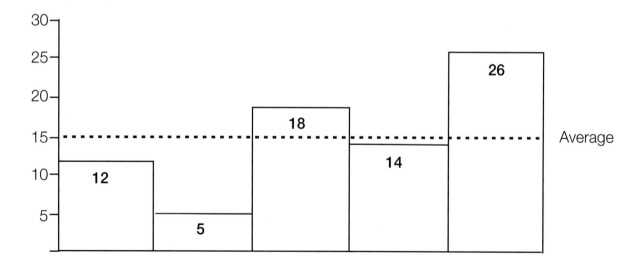

This kind of average is called the mean, or sometimes, mean average.

There are other ways of expressing the idea of typicality in a group of numbers: these are median and mode.

The median of a group of numbers is the middle number, when the numbers are arranged in ascending order of size.

For example, in the group **1, 3, 5, 7, 9,**   **5** is the median.

If there isn't a middle number, as in **3, 4, 6, 7,** you need to find the average (mean) of the two middle numbers. In this case, the two middle numbers are **4** and **6**, add these together **(4 + 6 = 10)**, then divide **10** by **2**, which gives the answer **5**, so **5** is the median.

The mode of a group of numbers is the number that occurs most often in the group.

For example, in the group **5, 6, 6, 7, 8,**   **6** is the mode.

There can be more than one mode in a group, for example, in the group **5, 6, 6, 7, 7, 8, 6** and **7** are the modes.

In everyday use, the term average is used in a general way, for example, average speed, average age, average height, etc.; median and mode are less commonly used than mean average.

Averages are often used to enable comparisons to be made, for example, you might be above average in height, earn less than the average wage and your journey to work might take an average of 30 minutes or an hour depending on whether you travel in the rush hour.

HD1/L2.3

Averages – continued

A footballer's goal average is the mean average number of goals he has scored in all his matches so far. If he has played 8 matches and scored 1, 0, 2, 3, 1, 0, 1, 2 goals in each match his goal average is:

$$1 + 0 + 2 + 3 + 1 + 0 + 1 + 2 = 10 = 1\tfrac{1}{4}$$

So, he scores, on average, $1\tfrac{1}{4}$ goals each match. Obviously, this could not be exactly true, but it does give us an idea of how good he is at scoring goals.

HD1/L2.3

Problems involving averages

1. A cricketer scores 51, 60, 44, 72, 46, 55 and 60 runs in 7 games. What is his batting average?

2. There are 6 classes in Yasmin's year at school with 32, 38, 41, 29, 40 and 36 children. What is the average size of a class?

3. I bought 6 boxes of matches, and noticed the boxes said 'average contents 48'. Having nothing better to do one day, I counted the matches in each box and there were 51, 49, 48, 49, 47 and 45. Was the number on the box right?

4. In the end of term exams, Chris got these marks: Maths 67, Science 60, History 70, French 58, English 72. What was his average exam mark?

5. What is the average speed of a car that travels 90 miles in 2 hours?

If you could travel on a car journey without slowing down or stopping, the speedometer would show the same figure all the time – you would be travelling at a constant speed. You might be able to do this on a motorway, but generally speed may vary from 70 mph down to 30 mph in towns or even 5 mph in slow heavy traffic. **Average speed** is one way of saying how fast we would be travelling if we could keep the same speed all the way.

An average speed of say 55 mph may mean that sometimes we were going much faster, say 75 mph, and sometimes as slow as 40 mph. A lower average speed of say 50 mph means that we were generally going more slowly, and so the journey would take longer.

To work out average speed: take the total distance travelled, and divide by the time taken.

For example:

A journey took 2 hours. The distance travelled was 120 miles. Find the average speed in miles per hour.

distance = 120 miles, time 2 hrs
average speed = 120 ÷ 2 = 60 mph

Find these average speeds in miles per hour:

1. A train travels 450 miles in 3 hours.

2. A cyclist travels 46 miles in 1 hour 30 minutes.

3. A plane flies 3,060 miles in 6 hours.

4. I'm going on a sponsored walk. If I can walk an average of 4 miles per hour, how long will it take me to walk the 21 mile round trip?

5. My 25 mile journey home usually takes 30 minutes by train, but in the rush hour by car it takes 50 minutes. By how much does my average speed vary according to whether I travel by train or by car?

6. A plane flies the 3000 miles across the Atlantic in 7 hours. What is its average speed? How long would the plane take to fly 4500 miles?

N1/L2.4

15 +
12
___

124 +
353
___

25 ×
18
___

205 +
176
___

21 ×
4
___

76 ×
6
___

153 ×
5
___

107 ×
12
___

24 ×
10
___

68 ×
9
___

84 ×
7
___

432 ×
8
___

52 −
21
___

562 −
42
___

258 −
136
___

30 −
18
___

243 −
27
___

160 −
53
___

3)636

4)824

6)642

8)248

3)195

2)6322

45 + 607 32	7653 + 8924 3799	876 + 4609 24

320 × 9	436 × 20	320 × 27

458 − 36	516 − 372	877 × 201

432 − 327	563 − 384	710 − 610

$6\overline{)4632}$　　　　$9\overline{)728}$　　　　$11\overline{)3467}$

$15\overline{)4862}$　　　　$26\overline{)8063}$　　　　$44\overline{)36254}$

A series is a group of numbers that have something in common with each other.

You can create a series of numbers by deciding what it is the series will have in common and then explore the patterns in the numbers that result.

Once you know a series, you can predict what the next numbers will be because they will continue the pattern.

The Fibonacci series

The Fibonacci series of numbers is **1, 1, 2, 3, 5, 8, 13, 21**… where each number (after the first two) is the sum of the two numbers immediately preceding it.

It's called the Fibonacci series after the man who identified it in the year 1202.

The Fibonacci numbers appear in many situations in nature. For example, the numbers of petals on flowers tend to be Fibonacci numbers, such as the iris, with 3 petals, and michaelmas daisies with 55 or 89 petals.

Continue the Fibonacci series and make up some series of your own.

Calculators take the effort out of adding up, taking away, multiplying and dividing.

Most calculators can work out percentages and square roots as well, and, generally, the more expensive a calculator is the more sophisticated it is likely to be, and the more operations it should be able to do.

When buying a calculator for your own use, decide what you are going to use it for. There's no point in having a calculator with lots of buttons on it that you never use.

A calculator will always give you the right answer if you know how to use it.

For example, suppose the calculation is **35 + 27**.

You press the button marked **3**, then the one marked **5**, then **+** then **2**, then **7**, then **=**. The number that comes up on the display will be the answer.

You do exactly the same to take away, multiply and divide. Press the buttons in the order of the digits in the numbers.

Try it with these:

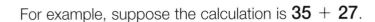

28 − 16 =	**306 ÷ 9 =**
19 × 5 =	**268 × 45 =**

You need to be able to do three things to get the best out of your calculator.

1. Know how to put a calculation into the calculator, in other words, which buttons to press and in what order.

2. See through to the maths in a problem you come across in daily life. In other words, you need to identify the calculation that needs to be done.

3. Be able to tell if the answer makes sense in the context.

See Book 4 for more on using a calculator or other aid to calculation.

N2/E3.4

How well can you add? p.3

3 + 4 = 7   6 + 6 = 12   8 + 7 = 15   5 + 9 = 14

3 +	9 +	7 +	8 +
4	8	6	8
7	2	3	3
14	1	4	2
	20	20	21

14 +	26 +	54 +	36 +
5	33	20	43
19	59	74	79

17 +	64 +	53 +	77 +
29	58	78	65
46	122	131	142

315 +	452 +	506 +	375 +
137	249	84	460
452	701	590	835

128 +	607 +	529 +	658 +
251	320	18	99
436	392	753	776
815	1319	1300	1533

532 +	7520 +	3479 +	2496 +
1807	165	1009	4032
321	3178	3795	1006
14	4221	4556	28
2674	15084	12839	7562

How to add, p.4

59 +	37 +	88 +	65 +
42	53	17	28
101	90	105	93

Adding single numbers, p.5

5 + 5 = 10   7 + 3 = 10   1 + 9 = 10   2 + 8 = 10
3 + 7 = 10   2 + 2 = 4   4 + 6 = 10   6 + 4 = 10
9 + 1 = 10   8 + 2 = 10   8 + 1 = 9   2 + 5 = 7
3 + 6 = 9   5 + 3 = 8   7 + 1 = 8   7 + 2 = 9
4 + 4 = 8   5 + 4 = 9   7 + 7 = 14   9 + 8 = 17
9 + 6 = 15   5 + 8 = 13   6 + 7 = 13   8 + 6 = 14
3 + 3 = 6   9 + 9 = 18   6 + 6 = 12   4 + 9 = 13
7 + 8 = 15   7 + 5 = 12

6 +	9 +	9 +	5 +	7 +
7	6	7	4	5
3	5	7	8	6
16	20	23	17	18

4 +	1 +	7 +	8 +	3 +
3	9	2	5	9
8	7	8	6	9
15	17	17	19	21

5 +	8 +	2 +	7 +	9 +
9	9	7	5	4
7	6	8	3	5
21	23	17	15	18

4 +	9 +	8 +	3 +	9 +
7	3	6	8	3
6	2	4	7	9
5	7	9	5	4
22	21	27	23	25

Adding double figures, p.6

TU	TU	TU	TU	TU
48 +	66 +	28 +	72 +	56 +
13	15	14	19	26
61	81	42	91	82

69 +	43 +	69 +	57 +	88 +
22	27	14	27	4
91	70	83	84	92

76 +	35 +	27 +	39 +	58 +
18	49	47	32	18
94	84	74	71	76

Adding-up problems, p.6

1. £4.90
2. Tuesday 15th of May
3. £125.70
4. Arsenal 2, West Ham 3
5. Jean is 31
6. £2.37
7. 13th June to 27th June
8. £2.65
9. Add 5 to the current year
10. Add 7 to today's date

Adding three-figure numbers, p.7

264 +	382 +	523 +
513	116	371
777	498	894

483 +	276 +	278 +
513	413	321
996	689	599

287 +	463 +	862 +
512	234	117
799	697	979

354 +	642 +	572 +
625	123	315
979	765	887

521 +	518 +	327 +
337	261	162
858	779	489

634 +	776 +	343 +
312	121	514
946	897	857

Answers

How well can you take away? p.8

$2 - 1 = 1 \qquad 6 - 4 = 2 \qquad 10 - 5 = 5 \qquad 8 - 2 = 6$
$11 - 3 = 8 \qquad 15 - 6 = 9$

60 −	95 −	99 −	16 −
32	87	7	13
28	8	92	3

80 −	100 −	250 −	327 −
78	99	125	129
2	1	125	198

472 −	588 −	536 −	371 −
365	298	156	203
107	290	380	168

318 −	407 −	894 −	243 −
89	316	776	75
229	91	118	168

3604 −	2060 −	58762 −	27908 −
1759	399	6898	8799
1845	1661	51864	19109

Take away – no borrowing, p.9

15 −	14 −	13 −	19 −
3	2	3	8
12	12	10	11

27 −	25 −	28 −	29 −
13	14	11	18
14	11	17	11

36 −	37 −	35 −	34 −
21	12	21	22
15	25	14	12

58 −	54 −	56 −	55 −
31	41	24	22
27	13	32	33

64 −	62 −	65 −	67 −
51	41	42	16
13	21	23	51

94 −	88 −	78 −
42	67	67
52	21	11

How to take away – borrowing method, p.10

$6^12 -$ or $6^12 -$	$4^11 -$ or $4^11 -$	$9^13 -$ or $9^13 -$	$8^14 -$ or $8^14 -$
^{137} $^4\cancel{3}7$	^{118} $^2\cancel{1}8$	^{125} $^3\cancel{2}5$	^{126} $^3\cancel{2}6$
25 25	23 23	68 68	58 58

How to take away – counting-on method, p.11

84 −	253 −	405 −	1562 −
76	166	291	944
8	87	114	618

How to take away – decomposition method, p.12, 13

85 −	42 −	254 −	526 −	− 373
48	17	149	218	127
37	25	105	308	246

204 −	508 −	301 −	205 −	502 −
158	99	253	88	425
46	409	48	117	77

Take away, p.14

61 −	43 −	72 −	55 −	37 −
35	28	49	47	19
26	15	23	8	18

24 −	81 −	62 −	45 −	172 −
16	38	17	38	89
8	43	45	7	83

423 −	833 −	602 −	510 −	304 −
76	189	363	297	148
347	644	239	213	156

Taking away problems, p.14

1. Rafiq was 32 14 years ago.
2. I had £7.02 left.
3. When it's 11 o'clock in the morning in Britain it's 6 am in New York.
4. (Work this out by taking 15 minutes away from the time now).
5. My Great Grandmother was 68 when she died.
6. This month is (…) 5 months ago it was (…).
7. Danny's sister is 41.
8. There are 19 women in the room.
9. Jamal earned £229 before his pay rise.
10. Leroy earns £14,081 above the national average.

How well can you multiply? p.15

$4 \times 5 = 20 \qquad 2 \times 6 = 12 \qquad 7 \times 8 = 56 \qquad 5 \times 9 = 45$
$6 \times 6 = 36 \qquad 12 \times 2 = 24$

18 ×	26 ×	33 ×	69 ×
3	4	5	6
54	104	165	414

20 ×	84 ×	62 ×	49 ×	98 ×
5	7	7	6	8
100	588	434	294	784

88 ×	104 ×	326 ×	230 ×	345 ×
6	4	7	6	9
528	416	2282	1380	3105

28 ×	35 ×	271 ×	198 ×	307 ×
10	30	11	11	12
280	1050	2981	2178	3684

450 ×	586 ×	349 ×	727 ×	89 ×
12	15	21	34	64
5400	8790	7329	24718	5696

Answers

608 ×	337 ×	750 ×	1834 ×	2072 ×
60	48	29	56	37
36480	16176	21750	102704	76664

158 ×	326 ×	872 ×	1760 ×	8121×
99	150	206	367	286
15642	48900	179632	645920	2322606

Fill in the blanks, p.17

(5 × 9 = 45) 3 × 7 = 21 9 × 10 = 90 6 × 5 = 30

7 × 6 = 42 (8 × 9 = 72) 6 × 8 = 48 4 × 9 = 36

(9 × 8 = 72) (9 × 7 = 63) (6 × 9 = 54) 8 × 8 = 64

(8 × 7 = 56) (7 × 8 = 56) (9 × 5 = 45) 4 × 3 = 12

5 × 4 = 20 (9 × 6 = 54)

8 × 8 = 64   7 × 8 = 56   9 × 4 = 36   9 × 9 = 81
5 × 9 = 45   5 × 7 = 35   7 × 9 = 63   7 × 7 = 49
5 × 8 = 40

6 × 10 = 60 or 10 × 6 = 60
3 × 7 = 21 or 7 × 3 = 21
2 × 9 = 18 or 9 × 2 = 18 or 3 × 6 = 18 or 6 × 3 = 18
2 × 12 = 24 or 12 × 2 = 24 or 4 × 6 = 24 or
6 × 4 = 24 or 3 × 8 = 24 or 8 × 3 = 24
4 × 4 = 16 or 2 × 8 = 16 or 8 × 2 = 16
3 × 6 = 18 or 6 × 3 = 18 or 2 × 9 = 18 or 9 × 2 = 18
4 × 6 = 24 or 6 × 4 = 24 or 2 × 12 = 24 or
12 × 2 = 24 or 8 × 3 = 24 or 3 × 8 = 24
2 × 8 = 16 or 8 × 2 = 16 or 4 × 4 = 16
3 × 11 = 33 or 11 × 3 = 33
7 × 5 = 35 or 5 × 7 = 35
6 × 9 = 54 or 9 × 6 = 54
6 × 8 = 48 or 8 × 6 = 48

Times tables, p.20

One times table: **p.20**

1 × 1 = 1   one times one is one
2 × 1 = 2
3 × 1 = 3
4 × 1 = 4
5 × 1 = 5
6 × 1 = 6
7 × 1 = 7
8 × 1 = 8
9 × 1 = 9
10 × 1 = 10
11 × 1 = 11
12 × 1 = 12

Two times table: **p.20**

1 × 2 = 2   one times two is two
2 × 2 = 4
3 × 2 = 6
4 × 2 = 8
5 × 2 = 10
6 × 2 = 12
7 × 2 = 14
8 × 2 = 16
9 × 2 = 18
10 × 2 = 20
11 × 2 = 22
12 × 2 = 24

Three times table: **p.20**

1 × 3 = 3   one times three is three
2 × 3 = 6
3 × 3 = 9
4 × 3 = 12
5 × 3 = 15
6 × 3 = 18
7 × 3 = 21
8 × 3 = 24
9 × 3 = 27
10 × 3 = 30
11 × 3 = 33
12 × 3 = 36

Four times table: **p.20**

1 × 4 = 4   one times four is four
2 × 4 = 8
3 × 4 = 12
4 × 4 = 16
5 × 4 = 20
6 × 4 = 24
7 × 4 = 28
8 × 4 = 32
9 × 4 = 36
10 × 4 = 40
11 × 4 = 44
12 × 4 = 48

Five times table: **p.21**

1 × 5 = 5   one times five is five
2 × 5 = 10
3 × 5 = 15
4 × 5 = 20
5 × 5 = 25
6 × 5 = 30
7 × 5 = 35
8 × 5 = 40
9 × 5 = 45
10 × 5 = 50
11 × 5 = 55
12 × 5 = 60

Six times table: **p.21**

$1 \times 6 = 6$ one times six is six
$2 \times 6 = 12$
$3 \times 6 = 18$
$4 \times 6 = 24$
$5 \times 6 = 30$
$6 \times 6 = 36$
$7 \times 6 = 42$
$8 \times 6 = 48$
$9 \times 6 = 54$
$10 \times 6 = 60$
$11 \times 6 = 66$
$12 \times 6 = 72$

Seven times table: **p.21**

$1 \times 7 = 7$ one times seven is seven
$2 \times 7 = 14$
$3 \times 7 = 21$
$4 \times 7 = 28$
$5 \times 7 = 35$
$6 \times 7 = 42$
$7 \times 7 = 49$
$8 \times 7 = 56$
$9 \times 7 = 63$
$10 \times 7 = 70$
$11 \times 7 = 77$
$12 \times 7 = 84$

Eight times table: **p.21**

$1 \times 8 = 8$ one times eight is eight
$2 \times 8 = 16$
$3 \times 8 = 24$
$4 \times 8 = 32$
$5 \times 8 = 40$
$6 \times 8 = 48$
$7 \times 8 = 56$
$8 \times 8 = 64$
$9 \times 8 = 72$
$10 \times 8 = 80$
$11 \times 8 = 88$
$12 \times 8 = 96$

Nine times table: **p.22**

$1 \times 9 = 9$ one times nine is nine
$2 \times 9 = 18$
$3 \times 9 = 27$
$4 \times 9 = 36$
$5 \times 9 = 45$
$6 \times 9 = 54$
$7 \times 9 = 63$
$8 \times 9 = 72$
$9 \times 9 = 81$
$10 \times 9 = 90$
$11 \times 9 = 99$
$12 \times 9 = 108$

Ten times table: **p.22**

$1 \times 10 = 10$ one times ten is ten
$2 \times 10 = 20$
$3 \times 10 = 30$
$4 \times 10 = 40$
$5 \times 10 = 50$
$6 \times 10 = 60$
$7 \times 10 = 70$
$8 \times 10 = 80$
$9 \times 10 = 90$
$10 \times 10 = 100$
$11 \times 10 = 110$
$12 \times 10 = 120$

Eleven times table: **p.22**

$1 \times 11 = 11$ one times eleven is eleven
$2 \times 11 = 22$
$3 \times 11 = 33$
$4 \times 11 = 44$
$5 \times 11 = 55$
$6 \times 11 = 66$
$7 \times 11 = 77$
$8 \times 11 = 88$
$9 \times 11 = 99$
$10 \times 11 = 110$
$11 \times 11 = 121$
$12 \times 11 = 132$

Twelve times table: **p.22**

$1 \times 12 = 12$ one times twelve is twelve
$2 \times 12 = 24$
$3 \times 12 = 36$
$4 \times 12 = 48$
$5 \times 12 = 60$
$6 \times 12 = 72$
$7 \times 12 = 84$
$8 \times 12 = 96$
$9 \times 12 = 108$
$10 \times 12 = 120$
$11 \times 12 = 132$
$12 \times 12 = 144$

Times table test sheet, p.24

$2 \times 4 = 8$	$7 \times 8 = 56$	$8 \times 9 = 72$
$3 \times 2 = 6$	$4 \times 10 = 40$	$7 \times 7 = 49$
$7 \times 3 = 21$	$6 \times 3 = 18$	$4 \times 9 = 36$
$2 \times 5 = 10$	$2 \times 9 = 18$	$7 \times 9 = 63$
$3 \times 10 = 30$	$7 \times 5 = 35$	$10 \times 9 = 90$
$5 \times 11 = 55$	$7 \times 12 = 84$	$11 \times 10 = 110$
$3 \times 3 = 9$	$10 \times 5 = 50$	$1 \times 10 = 10$
$4 \times 4 = 16$	$9 \times 9 = 81$	$4 \times 8 = 32$
$5 \times 4 = 20$	$5 \times 8 = 40$	$6 \times 9 = 54$
$6 \times 5 = 30$	$9 \times 6 = 54$	$2 \times 6 = 12$
$8 \times 4 = 32$	$6 \times 8 = 48$	$9 \times 3 = 27$
$7 \times 6 = 42$	$8 \times 2 = 16$	$8 \times 8 = 64$
$4 \times 7 = 28$	$8 \times 7 = 56$	$10 \times 6 = 60$
$6 \times 6 = 36$	$5 \times 9 = 45$	$8 \times 10 = 80$
$5 \times 5 = 25$	$10 \times 7 = 70$	$4 \times 12 = 48$

Short multiplication, p.26

42×	53×	67×	35×	84×
6	4	5	7	9
252	212	335	245	756

63×	21×	70×	97×	123×
6	8	3	6	6
378	168	210	582	738

518×	704×	806×	902×	510×
9	7	5	7	8
4662	4928	4030	6314	4080

Short multiplication problems, p.26

1. They paid £33.00 altogether.
2. (Multiply 10 times the cost of one stamp.)
3. My return fare to work costs me £18.75 for 5 days.
4. 6 shrubs cost £35.94.
5. There are 26 flats in the block.
6. I'll have £75.
7. It will cost £40.00 to tile the bathroom.
8. 6 litres cost £4.80.
9. The price charged was not correct. It should have been £7.05.

Multiplying by 10, 20, 30 and so on, p.28

15×	15×	15×	15×	15×	15×
10	20	30	40	50	60
150	300	450	600	750	900

23×	23×	23×	23×	23×	23×
10	20	30	40	50	60
230	460	690	920	1150	1380

31×	31×	31×	31×	31×	31×
10	30	50	70	80	90
310	930	1550	2170	2480	2790

42×	42×	42×	42×	42×	42×
20	40	50	60	70	90
840	1680	2100	2520	2940	3780

56×	56×	56×	56×	56×	56×
10	30	40	60	70	80
560	1680	2240	3360	3920	4480

67×	67×	67×	67×	67×	67×
10	20	30	40	50	60
670	1340	2010	2680	3350	4020

How to do long multiplication, p.30

65×	35×	79×	58×	86×	44×
24	47	35	19	34	29
1560	1645	2765	1102	2924	1276

Long multiplication, p.30

66×	38×	79×	87×	94×
27	45	4	26	35
1782	1710	316	2262	3290

53×	29×	71×	68×	77×
17	46	32	51	36
901	1334	2272	3468	2772

89×	62×	304×	210×	781×
25	28	31	16	45
2225	1736	9424	3360	35145

Long multiplication problems, p.31

1. She earns £4680 a year.
2. No. 10 feet by 12 feet is 120 square feet.
3. I would save £9 by buying in bulk.
4. 20 cigarettes a day is 7300 in a year.
5. Alun bought 180 plants altogether.
6. Netta earned £825 altogether.
7. Dave won the bet. There were only 288 holes in the piece of pegboard.
8. I pay £7280 rent in a year.
9. The washing machine would cost £528 on Hire Purchase.
10. I need 20 square metres which would costs £379. You can approximate to £380 by saying 20 × £19.

Factors, factorising and prime numbers, p.31

Find the factors of these numbers:
18: 2 and 9; 2 and 3 and 3;
25: 5 and 5
9: 3 x 3
14: 2 and 7
21: 3 and 7

Prime numbers, p.32

2, 3, 5, 7, 11, 13, 17, 19, 21, 29, 31, 37, 41, 43, 47, 51, 53, 57, 59, 61, 67, 71, 73, 79, 83, 87, 91, 97

Prime factors, p.32

Break down these numbers into prime factors:
$16 = 4 \times 4 = 2 \times 2 \times 2 \times 2$
$12 = 2 \times 6 = 2 \times 2 \times 3 \quad 3 \times 4 = 3 \times 2 \times 2$
$18 = 3 \times 6 = 3 \times 2 \times 3 \quad 2 \times 9 = 2 \times 3 \times 3$
$21 = 3 \times 7$
$30 = 3 \times 10 = 3 \times 2 \times 5 \quad 6 \times 5 = 2 \times 3 \times 5$
$27 = 3 \times 9 = 3 \times 3 \times 3$
$24 = 4 \times 6 = 2 \times 2 \times 2 \times 3 \quad 3 \times 8 = 3 \times 2 \times 2 \times 2$
$25 = 5 \times 5$
$36 = 6 \times 6 = 2 \times 3 \times 2 \times 3 \quad 4 \times 9 = 2 \times 2 \times 3 \times 3$

Multiplying and dividing, p.33

$12 = 2 \times 6$; $12 = 6 \times 2$; $12 \div 6 = 2$; $12 \div 2 = 6$;
$12 = 3 \times 4$; $12 = 4 \times 3$; $12 \div 4 = 3$; $12 \div 3 = 4$;
$14 = 2 \times 7$; $14 = 7 \times 2$; $14 \div 7 = 2$; $14 \div 2 = 7$;
$18 = 2 \times 9$; $18 = 9 \times 2$; $18 \div 9 = 2$; $18 \div 2 = 9$;
$18 = 3 \times 6$; $18 = 6 \times 3$; $18 \div 6 = 3$; $18 \div 3 = 6$;
$24 = 6 \times 4$; $24 = 4 \times 6$; $24 \div 6 = 4$; $24 \div 4 = 6$;
$24 = 3 \times 8$; $24 = 8 \times 3$; $24 \div 3 = 8$; $24 \div 8 = 3$;
$24 = 2 \times 12$; $24 = 12 \times 2$; $24 \div 12 = 2$; $24 \div 2 = 12$;
$56 = 4 \times 14$; $56 = 14 \times 4$; $56 \div 4 = 14$; $56 \div 14 = 4$;
$56 = 2 \times 28$; $56 = 28 \times 2$; $56 \div 28 = 2$; $56 \div 2 = 28$;
$56 = 7 \times 8$; $56 = 8 \times 7$; $56 \div 7 = 8$; $56 \div 8 = 7$;
$35 = 5 \times 7$; $35 = 7 \times 5$; $35 \div 5 = 7$; $35 \div 7 = 5$;
$60 = 6 \times 10$; $60 = 10 \times 6$; $60 \div 10 = 6$; $60 \div 6 = 10$;
$33 = 3 \times 11$; $33 = 11 \times 3$; $33 \div 11 = 3$; $33 \div 3 = 11$.

Powers and indices, p.34

$6^3 = 6 \times 6 \times 6$
$5^4 = 5 \times 5 \times 5 \times 5$
$2^3 = 2 \times 2 \times 2$
$9^8 = 9 \times 9 \times 9 \times 9 \times 9 \times 9 \times 9 \times 9$
$7^2 = 7 \times 7$
$8^5 = 8 \times 8 \times 8 \times 8 \times 8$
$3^3 = 3 \times 3 \times 3$
$4^2 = 4 \times 4$
$15^3 = 15 \times 15 \times 15$
$12^{10} = 12 \times 12 \times 12 \times 12 \times 12 \times 12 \times 12 \times 12 \times 12 \times 12$

How well can you divide, p.34

$10 \div 2 = 5$; $15 \div 3 = 5$; $16 \div 4 = 4$; $20 \div 5 = 4$;
$66 \div 6 = 11$

$2 \overline{)26} = 13$ $3 \overline{)636} = 212$ $4 \overline{)408} = 102$ $5 \overline{)550} = 110$

$3 \overline{)999} = 333$ $4 \overline{)250} = 62.5$ $3 \overline{)810} = 270$ $8 \overline{)656} = 82$

$8 \overline{)306} = 38.25$ $4 \overline{)7790} = 1947.5$ $20 \overline{)76} = 3.8$ $23 \overline{)46} = 2$

$18 \overline{)360} = 20$ $12 \overline{)993} = 82.75$ $12 \overline{)376} = 31.3$ recurring or $31\tfrac{1}{3}$

Writing division sums, p.35

$12 \div 4 = 3$ $\dfrac{12}{4} = 3$ $4 \overline{)12} = 3$

$6 \overline{)42} = 7$ $42 \div 6 = 7$ $\dfrac{42}{6} = 7$

$\dfrac{18}{9} = 2$ $18 \div 9 = 2$ $2 \overline{)18} = 9$

$36 \div 9 = 4$ $\dfrac{36}{9} = 4$ $9 \overline{)36} = 4$

Short division, p.35

$15 \div 3 = 5$
$10 \div 2 = 5$
$9 \div 3 = 3$
$14 \div 2 = 7$
$20 \div 4 = 5$

Fill in the blanks, p.37

$12 \div 3 = 4$ $24 \div 6 = 4$ $18 \div 9 = 2$ $27 \div 3 = 9$
$15 \div 5 = 3$ $18 \div 6 = 3$ $25 \div 5 = 5$ $36 \div 6 = 6$
$42 \div 7 = 6$ $63 \div 9 = 7$ $100 \div 10 = 10$ $88 \div 8 = 11$

Divide, p.38

$4 \overline{)284} = 71$ $5 \overline{)450} = 90$ $6 \overline{)426} = 71$ $7 \overline{)560} = 80$ $4 \overline{)3916} = 979$

$6 \overline{)600} = 100$ $9 \overline{)819} = 91$ $3 \overline{)153} = 51$ $5 \overline{)205} = 41$ $3 \overline{)450} = 150$

$2 \overline{)184} = 92$ $3 \overline{)999} = 333$ $2 \overline{)246} = 123$ $7 \overline{)497} = 71$ $7 \overline{)6342} = 906$

$2 \overline{)868} = 434$ $6 \overline{)4812} = 802$ $9 \overline{)7218} = 802$ $7 \overline{)553} = 79$ $5 \overline{)732} = 146.4$

Short division problems, p.38

1. We each paid £20.
2. £1.26 works out as 42p per pack, so it is cheaper to buy a 3-carton pack rather than three separate cartons.
3. Jagdish needs to save for 29 months.
4. Ming-Li can make 2 shelves out of the plank.
5. It will take Bob 5 weeks to lose 10 kilograms.
6. Each child will have 2 bars of chocolate and there will be 4 bars left over.
7. 10 records can be played during the programme.
8. George drove 20 miles a day on average.
9. The new flat costs £18 a month more.
10. Each bar of soap costs 40p.

Long division problems, p.40

1. There are 16 seats in each row.
2. No. There will be enough for 4 rolls each.
3. There are 24 apples in each case.
4. 17 chocolates will pack into 15 boxes with 1 left over.
5. I'll need 3 boxes.
6. If I order 14 crates I'll have 504 bottles of beer.
7. I need to order 13 coaches but there will be 20 spare seats.
8. I earn £282.31 per week.
9. I need to save £37.50 each month.

Problems involving averages, p.44

1. The cricketer's batting average is 55.43.
2. The average size of a class is 36 children.
3. The average contents are just over 48.
4. Chris's average exam mark was 65.4.
5. The average speed is 45 miles per hour.

Average speeds, p.45

1. The train travels at 150 miles per hour.
2. The cyclist travels at just over 30 miles per hour (30.67mph).
3. The plane flies at 510 miles per hour.
4. It will take me 5 and a quarter hours to do the sponsored walk.
5. My average speed is 50 mph by train and 30 mph by car, so my average speed varies by 20 mph.
6. The plane's average speed is 429 mph. It would take the plane 10 and a half hours to fly 4500 miles.

Mixed sums 1, p.46

15+	124+	25×	205+
12	353	18	176
27	477	450	381

21×	76×	153×	107×
4	6	5	12
84	456	765	1284

24×	68×	84×	432×
10	9	7	8
240	612	588	3456

52−	562−	258−	30−
21	42	136	18
31	520	122	12

243−	160−	$3\overline{)636}$ 212	$4\overline{)824}$ 206
27	53		
216	107		

$6\overline{)642}$ 107	$8\overline{)248}$ 31	$3\overline{)195}$ 65	$2\overline{)6322}$ 3161

Mixed sums 2, p.47

45+	7653+	876+
607	8924	4609
32	3799	24
684	20376	5509

320×	436×	320×
9	20	27
2880	8720	8640

458−	516−	877×
36	372	201
422	144	176277

432−	563−	710−
327	384	610
105	179	100

$6\overline{)4632}$ 772	$9\overline{)728}$ 80 rem 8	$11\overline{)3467}$ 315 rem 2

$15\overline{)4862}$ 324 rem 2	$26\overline{)8063}$ 310 rem 3	$44\overline{)36254}$ 823 rem 42 or 823.95

Using the calculator, p.48

$28 - 16 = 12$ $306 \div 9 = 34$

$19 \times 5 = 95$ $268 \times 45 = 12{,}060$

Contents

Introduction

This book covers fractions, decimals, percentages, ratio and proportion: what they mean, how to do calculations with them and how they relate to each other.

It aims to help you to build on your knowledge and experience and to make connections between different aspects of mathematics.

As in the rest of the pack, this book aims to help you to develop what the educationalist Richard Skemp[1] called 'relational understanding' (knowing what to do and why).

This is because we believe that relational mathematics is more adaptable to new tasks and easier to remember – and that you have a right to know what to do and why.

---

1. Skemp, R.R. (1978) 'Relational understanding and instrumental understanding', in *The Arithmetic Teacher,* 26, 3, pp 9–15

Fractions

Fractions are parts of one. You could think of fractions as what happens when a whole one is broken into pieces.

The bottom number (the denominator) tells you how many pieces the whole one is broken up into, and the top number (the numerator) tells you how many of those pieces you are concerned with in each case.

For example, $\frac{1}{2}$ means one split into two equal pieces: a half is one of those pieces.

You could also think of it as 1 divided by 2, or 1 divided into 2 equal pieces.

You can also write any division calculation like this. The number being divided goes on top and the number you're dividing by goes under the line at the bottom. In that case, the top number is called the dividend and the bottom number is called the divisor.

The division sign ÷ is an abstract representation of this, with dots standing in for the divisor and the dividend.

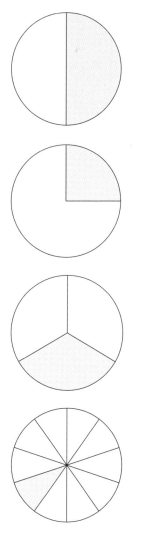 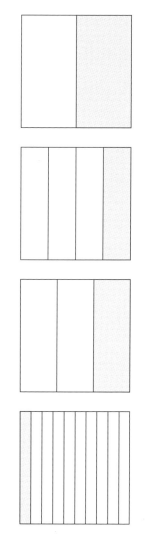 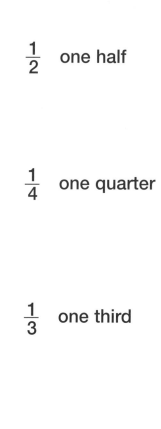

$\frac{1}{2}$ one half

$\frac{1}{4}$ one quarter

$\frac{1}{3}$ one third

$\frac{1}{10}$ one tenth

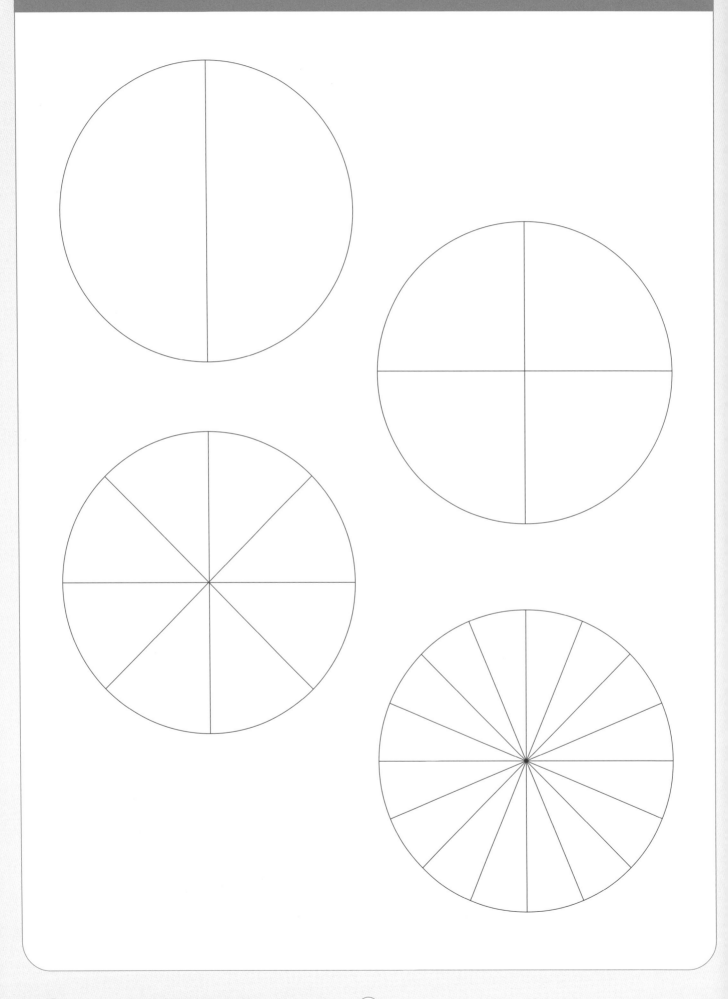

Shade in the fractions

Shade in these fractions.

$\dfrac{1}{2}$

$\dfrac{1}{2}$	$\dfrac{1}{2}$

$\dfrac{3}{10}$

$\dfrac{1}{10}$	$\dfrac{1}{10}$	$\dfrac{1}{10}$	$\dfrac{1}{10}$	$\dfrac{1}{10}$	$\dfrac{1}{10}$	$\dfrac{1}{10}$	$\dfrac{1}{10}$	$\dfrac{1}{10}$	$\dfrac{1}{10}$

$\dfrac{3}{4}$

$\dfrac{1}{4}$	$\dfrac{1}{4}$	$\dfrac{1}{4}$	$\dfrac{1}{4}$

$\dfrac{5}{6}$

$\dfrac{1}{6}$	$\dfrac{1}{6}$	$\dfrac{1}{6}$	$\dfrac{1}{6}$	$\dfrac{1}{6}$	$\dfrac{1}{6}$

$\dfrac{2}{5}$

$\dfrac{1}{5}$	$\dfrac{1}{5}$	$\dfrac{1}{5}$	$\dfrac{1}{5}$	$\dfrac{1}{5}$

$\dfrac{5}{12}$

$\dfrac{1}{12}$	$\dfrac{1}{12}$	$\dfrac{1}{12}$	$\dfrac{1}{12}$	$\dfrac{1}{12}$	$\dfrac{1}{12}$	$\dfrac{1}{12}$	$\dfrac{1}{12}$	$\dfrac{1}{12}$	$\dfrac{1}{12}$	$\dfrac{1}{12}$	$\dfrac{1}{12}$

$\dfrac{2}{3}$

$\dfrac{1}{3}$	$\dfrac{1}{3}$	$\dfrac{1}{3}$

$\dfrac{1}{5}$

$\dfrac{1}{5}$	$\dfrac{1}{5}$	$\dfrac{1}{5}$	$\dfrac{1}{5}$	$\dfrac{1}{5}$

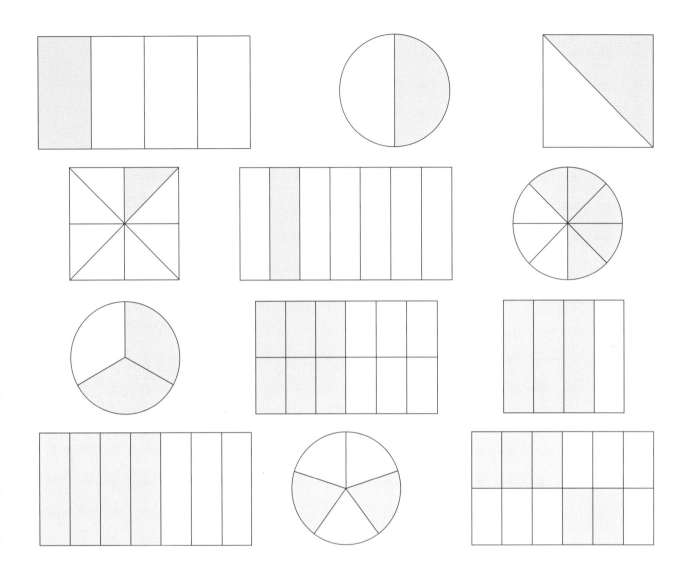

Shade in the fractions given.

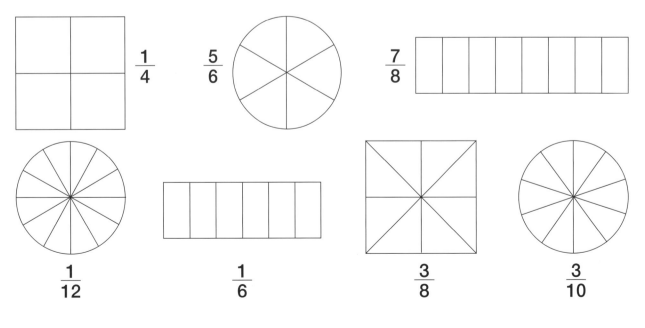

$\dfrac{1}{4}$ $\dfrac{5}{6}$ $\dfrac{7}{8}$

$\dfrac{1}{12}$ $\dfrac{1}{6}$ $\dfrac{3}{8}$ $\dfrac{3}{10}$

One whole one								one unit	1
1									

| One half | | | | One half | | | | two halves | $\frac{2}{2}$ |
| $\frac{1}{2}$ | | | | $\frac{1}{2}$ | | | | | |

| One quarter | | One quarter | | One quarter | | One quarter | | four quarters | $\frac{4}{4}$ |
| $\frac{1}{4}$ | | $\frac{1}{4}$ | | $\frac{1}{4}$ | | $\frac{1}{4}$ | | | |

| One eighth | One eighth | One eighth | One eighth | One eighth | One eighth | One eighth | One eighth | eight eighths | $\frac{8}{8}$ |
| $\frac{1}{8}$ | $\frac{1}{8}$ | $\frac{1}{8}$ | $\frac{1}{8}$ | $\frac{1}{8}$ | $\frac{1}{8}$ | $\frac{1}{8}$ | $\frac{1}{8}$ | | |

What is half of one?

What is half of a half?

What is half of a quarter?

What is a quarter of a half?

How many eighths in one?

How many eighths in a half?

How many quarters in one?

How many quarters in a half?

Fraction block $1 \frac{1}{3} \frac{1}{6} \frac{1}{12}$

One whole one		
1		

one unit   **1**

One third $\frac{1}{3}$	One third $\frac{1}{3}$	One third $\frac{1}{3}$

three thirds $\frac{3}{3}$

One sixth $\frac{1}{6}$	One sixth $\frac{1}{6}$	One sixth $\frac{1}{6}$	One sixth $\frac{1}{6}$	One sixth $\frac{1}{6}$	One sixth $\frac{1}{6}$

six sixths $\frac{6}{6}$

One twelfth $\frac{1}{12}$	One twelfth $\frac{1}{12}$	One twelfth $\frac{1}{12}$	One twelfth $\frac{1}{12}$	One twelfth $\frac{1}{12}$	One twelfth $\frac{1}{12}$	One twelfth $\frac{1}{12}$	One twelfth $\frac{1}{12}$	One twelfth $\frac{1}{12}$	One twelfth $\frac{1}{12}$	One twelfth $\frac{1}{12}$	One twelfth $\frac{1}{12}$

twelve twelfths $\frac{12}{12}$

What is a third of one?

What is half of a third?

What is half of a sixth?

How many sixths in one?

How many sixths in a third?

How many twelfths in a third?

How many twelfths in two thirds?

How many twelfths in one?

One whole one										one unit	1
1											
One fifth $\frac{1}{5}$		One fifth $\frac{1}{5}$		One fifth $\frac{1}{5}$		One fifth $\frac{1}{5}$		One fifth $\frac{1}{5}$		five fifths	$\frac{5}{5}$
One tenth $\frac{1}{10}$	One tenth $\frac{1}{10}$	One tenth $\frac{1}{10}$	One tenth $\frac{1}{10}$	One tenth $\frac{1}{10}$	One tenth $\frac{1}{10}$	One tenth $\frac{1}{10}$	One tenth $\frac{1}{10}$	One tenth $\frac{1}{10}$	One tenth $\frac{1}{10}$	ten tenths	$\frac{10}{10}$

What is a fifth of one?

What is half of a fifth?

What is ten tenths?

How many fifths in one?

How many tenths in a fifth?

How many tenths in three fifths?

How many tenths in one?

One whole

$\frac{2}{2}$ = 2 halves

$\frac{4}{4}$ = 4 quarters

$\frac{8}{8}$ = 8 eighths

$\frac{16}{16}$ = 16 sixteenths

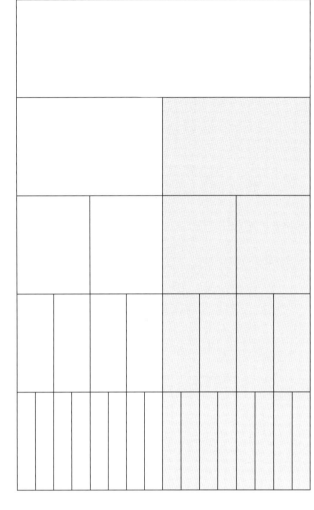

$\frac{1}{2}$

$\frac{2}{4}$ = $\frac{1}{2}$

$\frac{4}{8}$ = $\frac{2}{4}$ = $\frac{1}{2}$

$\frac{8}{16}$ = $\frac{4}{8}$ = $\frac{2}{4}$ = $\frac{1}{2}$

Complete these:

$\frac{1}{4}$ = $\frac{}{8}$ = $\frac{}{16}$

$\frac{3}{4}$ = $\frac{}{8}$ = $\frac{}{16}$

$\frac{1}{2}$ = $\frac{}{4}$ = $\frac{}{8}$ = $\frac{}{16}$

One whole

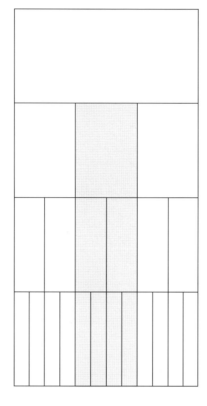

$\frac{3}{3}$ = 3 thirds

$\frac{1}{3}$

$\frac{6}{6}$ = 6 sixths

$\frac{2}{6}$ = $\frac{1}{3}$

$\frac{12}{12}$ = 12 twelfths

$\frac{4}{12}$ = $\frac{2}{6}$ = $\frac{1}{3}$

Complete these:　$\frac{1}{6}$ = $\frac{}{12}$　　　$\frac{5}{6}$ = $\frac{}{12}$　　　$\frac{2}{3}$ = $\frac{}{6}$ = $\frac{}{12}$

One whole

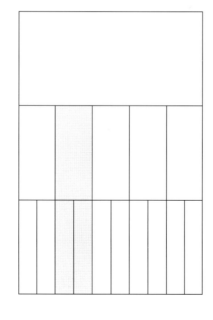

$\frac{5}{5}$ = 5 fifths

$\frac{1}{5}$

$\frac{10}{10}$ = 10 tenths

$\frac{2}{10}$ = $\frac{1}{5}$

Complete these:　$\frac{2}{5}$ = $\frac{}{10}$　　　$\frac{3}{5}$ = $\frac{}{10}$　　　$\frac{8}{10}$ = $\frac{}{5}$　　　$\frac{7}{5}$ = $\frac{}{10}$

N2/E3.1, N2/E3.2

Fraction block

Fractions with different denominators relate to each other as shown on the fraction block below.

Some line up with each other and some do not. Can you explain why?

One whole **1**

| $\frac{1}{2}$ | |

| $\frac{1}{3}$ | | |

| $\frac{1}{4}$ | | | |

| $\frac{1}{5}$ | | | | |

| $\frac{1}{8}$ | | | | | | | |

| $\frac{1}{10}$ | | | | | | | | | |

Fractions number line

$2\frac{1}{2}$ —— $\frac{5}{2}$

2 —— $\frac{4}{2}$

$1\frac{1}{2}$ —— $\frac{3}{2}$

1 —— $\frac{2}{2}$

$\frac{1}{2}$ —— $\frac{1}{2}$

0 —— 0

Halves

$2\frac{1}{4}$ —— $\frac{9}{4}$

2 —— $\frac{8}{4}$

$1\frac{3}{4}$ —— $\frac{7}{4}$

$1\frac{1}{2}$ —— $\frac{6}{4}$

$1\frac{1}{4}$ —— $\frac{5}{4}$

1 —— $\frac{4}{4}$

$\frac{3}{4}$ —— $\frac{3}{4}$

$\frac{1}{2}$ —— $\frac{2}{4}$

$\frac{1}{4}$ —— $\frac{1}{4}$

0 —— 0

Quarters

$2\frac{1}{6}$ —— $\frac{13}{6}$

2 —— $\frac{12}{6}$

$1\frac{5}{6}$ —— $\frac{11}{6}$

$1\frac{2}{3}$ —— $\frac{10}{6}$

$1\frac{1}{2}$ —— $\frac{9}{6}$

$1\frac{1}{3}$ —— $\frac{8}{6}$

$1\frac{1}{6}$ —— $\frac{7}{6}$

1 —— $\frac{6}{6}$

$\frac{5}{6}$ —— $\frac{5}{6}$

$\frac{2}{3}$ —— $\frac{4}{6}$

$\frac{1}{2}$ —— $\frac{3}{6}$

$\frac{1}{3}$ —— $\frac{2}{6}$

$\frac{1}{6}$ —— $\frac{1}{6}$

0 —— 0

Sixths

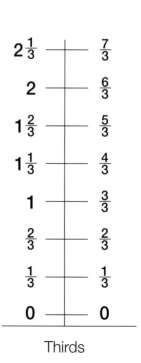

$2\frac{1}{3}$ —— $\frac{7}{3}$

2 —— $\frac{6}{3}$

$1\frac{2}{3}$ —— $\frac{5}{3}$

$1\frac{1}{3}$ —— $\frac{4}{3}$

1 —— $\frac{3}{3}$

$\frac{2}{3}$ —— $\frac{2}{3}$

$\frac{1}{3}$ —— $\frac{1}{3}$

0 —— 0

Thirds

N2/E3.2, N2/L1.1

1. $\frac{1}{4} \div 5 =$

2. $\frac{1}{2} + \frac{2}{3} - \frac{3}{4} =$

3. $\frac{1}{2} \times \frac{1}{3} =$

4. $\frac{5}{8} \div \frac{2}{3} =$

5. $3 - 1\frac{1}{9} =$

6. $\frac{2}{3} \times 2 =$

7. $1 - \frac{1}{4} =$

8. $3\frac{1}{3} + \frac{2}{3} =$

9. $2 + \frac{2}{5} =$

10. $1\frac{1}{2} \div 3\frac{3}{4} =$

11. $\frac{15}{16} \div \frac{6}{10} =$

12. $1 + \frac{1}{2} =$

13. $\frac{1}{6} \times 5 =$

14. $2\frac{1}{3} + 1\frac{5}{6} + 1\frac{1}{3} =$

To add together fractions which have the same denominator just add the top numbers.

For example: $\frac{3}{5} + \frac{1}{5} = \frac{4}{5}$ $\frac{1}{8} + \frac{3}{8} = \frac{4}{8} = \frac{1}{2}$

Try these: $\frac{1}{4} + \frac{1}{4}$ $\frac{2}{7} + \frac{4}{7}$ $\frac{3}{10} + \frac{7}{10}$ $\frac{1}{9} + \frac{1}{9}$

If the denominators are different, change one or both fractions so the bottom numbers are the same (this is called finding the common denominator), and then add the top numbers (the numerators).

For example: $\frac{1}{4} + \frac{3}{8} = \frac{2 + 3}{8} = \frac{5}{8}$

Here, we have changed $\frac{1}{4}$ to $\frac{2}{8}$ so we can add them together and get $\frac{5}{8}$

Another example: $\frac{1}{4} + \frac{2}{3} = \frac{3 + 8}{12} = \frac{11}{12}$

Here we have to change both quarters and thirds to twelfths before they can be added, 12 is the common denominator. It is the smallest number we can use in this case which is divisible by both 3 and 4. This is how we change the fractions:

What do you need to multiply 4 by to get 12? Answer 3, so multiply the top line by 3.

So $\frac{1}{4} = \frac{3}{12}$

Similarly, what do you multiply 3 by to get 12? Answer 4, so multiply the top line by 4.

$$\frac{2}{3} = \frac{8}{12}$$

Now try these:

$\frac{2}{3} + \frac{1}{6}$ $\frac{2}{3} + \frac{1}{2}$ $\frac{3}{4} + \frac{5}{8}$ $\frac{1}{5} + \frac{2}{3}$ $\frac{5}{6} + \frac{1}{9}$

Whole numbers and fractions (mixed numbers)

$$2\frac{3}{4} + 3\frac{1}{2} = \frac{11}{4} + \frac{7}{2} = \frac{11 + 14}{4} = \frac{25}{4} = 6\frac{1}{4}$$

When we have whole numbers and fractions together, before we can add, we change these mixed numbers into improper fractions. In improper fractions the top number (the numerator) is bigger than the bottom number (the denominator).

So $2\frac{3}{4} = \frac{11}{4}$ and $3\frac{1}{2} = \frac{7}{2}$ Then find the common denominator and add.

Try these:

$1\frac{1}{4} + 3\frac{1}{3}$ $2\frac{2}{5} + 4\frac{3}{10}$ $1\frac{3}{8} + 2\frac{1}{4}$ $6\frac{2}{3} + 2\frac{1}{2}$

N2/L2.4

Like adding, you can take away fractions which have the same number at the bottom (a common denominator).

When the denominators are already the same, just take away the top numbers.

For example: $\dfrac{3}{5} - \dfrac{1}{5} = \dfrac{2}{5}$ $\qquad\qquad \dfrac{3}{8} - \dfrac{1}{8} = \dfrac{2}{8} = \dfrac{1}{4}$

Try these: $\dfrac{3}{4} - \dfrac{1}{4}$ $\qquad \dfrac{5}{8} - \dfrac{3}{8}$ $\qquad \dfrac{6}{7} - \dfrac{1}{7}$ $\qquad \dfrac{2}{3} - \dfrac{1}{3}$ $\qquad \dfrac{9}{10} - \dfrac{3}{10}$

When fractions are different: If the denominators are different, we must change one or both fractions so that the bottom numbers are the same (i.e. find the common denominator), and then take away.

For example: $\dfrac{3}{8} - \dfrac{1}{4} = \dfrac{3-2}{8} = \dfrac{1}{8}$

Here, we have changed one of the fractions, $\dfrac{1}{4}$ to $\dfrac{2}{8}$ so we can take it away from $\dfrac{3}{8}$

Another example: $\dfrac{2}{3} - \dfrac{1}{4} = \dfrac{8-3}{12} = \dfrac{5}{12}$

Here, as for adding up, we have to change both thirds and quarters to twelfths (the common denominator) before we can take away.

Now try these:

$\dfrac{5}{6} - \dfrac{1}{3}$ $\qquad \dfrac{2}{3} - \dfrac{1}{2}$ $\qquad \dfrac{3}{4} - \dfrac{5}{8}$ $\qquad \dfrac{3}{5} - \dfrac{1}{3}$ $\qquad \dfrac{3}{4} - \dfrac{2}{5}$

Whole numbers and fractions (mixed numbers)

$$3\dfrac{1}{2} - 1\dfrac{3}{4} = \dfrac{7}{2} - \dfrac{7}{4} = \dfrac{14-7}{4} = \dfrac{7}{4} = 1\dfrac{3}{4}$$

Before we can take away, we must change the mixed numbers into improper fractions (so that the top number is bigger than the bottom number).

$3\dfrac{1}{2} = \dfrac{7}{2}$ (seven halves) $\qquad\qquad 1\dfrac{3}{4} = \dfrac{7}{4}$ (seven quarters)

Then find the common denominator and take away.

Try these:

$2\dfrac{1}{4} - 1\dfrac{3}{4}$ $\qquad\qquad 4\dfrac{1}{8} - 2\dfrac{1}{4}$ $\qquad\qquad 3\dfrac{1}{7} - 1\dfrac{3}{14}$ $\qquad\qquad 6\dfrac{3}{10} - 2\dfrac{4}{5}$

How to multiply fractions

Multiplying fractions is easy:

$\frac{1}{2} \times \frac{1}{2} = \frac{1}{4}$

Do the top line first.

Say $\mathbf{1 \times 1 = 1}$

Then do the bottom line.

Say: $\mathbf{2 \times 2 = 4}$

Answer: $\frac{1}{4}$ a quarter

The multiplication sign (x) is another way of saying 'of' so you could say:

$\frac{1}{2} \times \frac{1}{2} = \frac{1}{4}$ a half of a half is a quarter

Now try these:

$\frac{1}{2} \times \frac{1}{4} =$ $\qquad$ $\frac{1}{2} \times \frac{1}{8} =$

$\frac{1}{2} \times \frac{1}{6} =$ $\qquad$ $\frac{1}{2} \times \frac{1}{3} =$

$\frac{1}{3} \times \frac{1}{4} =$ $\qquad$ $\frac{1}{5} \times \frac{1}{3} =$

Multiplying fractions

$\frac{2}{3} \times \frac{3}{5} =$ $\qquad$ $\frac{4}{5} \times \frac{3}{4} =$ $\qquad$ $\frac{2}{3} \times \frac{3}{4} =$

$\frac{1}{4} \times \frac{7}{8} =$ $\qquad$ $\frac{7}{8} \times \frac{4}{9} =$ $\qquad$ $\frac{9}{10} \times \frac{2}{5} =$

$\frac{3}{5} \times \frac{1}{3} =$ $\qquad$ $\frac{2}{9} \times \frac{1}{2} =$ $\qquad$ $\frac{3}{7} \times \frac{1}{5} =$

$\frac{4}{9} \times \frac{1}{3} =$ $\qquad$ $\frac{1}{10} \times \frac{3}{16} =$ $\qquad$ $\frac{4}{13} \times \frac{2}{3} =$

How to divide fractions

Once you can multiply fractions, dividing them is easy:

$$\frac{1}{4} \div \frac{1}{3} =$$

Before you do anything else change ÷ to × and turn $\frac{1}{3}$ upside down to give you $\frac{3}{1}$.

When you change $\frac{1}{3}$ into $\frac{3}{1}$ the sum looks like this:

$$\frac{1}{4} \times \frac{3}{1} =$$

Now you can do it as a multiplication sum.

Do the top line first: $1 \times 3 = 3$

Then the bottom line: $4 \times 1 = 4$

Answer: $\frac{3}{4}$ three quarters

This may seem strange. To understand why it works, look at the first page in this book, which explains that a fraction is a way of showing the relationship between two numbers where the top number is divided by the bottom number. By turning $\frac{1}{3}$ upside down and changing the operation from division to multiplication, you're maintaining the relationship between the two fractions in the original sum ($\frac{1}{4} \div \frac{1}{3}$).

Now try these:

$$\frac{1}{5} \div \frac{1}{3} = \qquad \frac{1}{6} \div \frac{1}{5} =$$

$$\frac{1}{3} \div \frac{2}{9} = \qquad \frac{1}{9} \div \frac{1}{2} =$$

$$\frac{2}{7} \div \frac{1}{2} = \qquad \frac{1}{2} \div \frac{5}{7} =$$

Dividing fractions

$$\frac{3}{5} \div \frac{1}{2} = \qquad \frac{1}{9} \div \frac{3}{4} = \qquad \frac{3}{4} \div \frac{2}{3} = \qquad \frac{8}{9} \div \frac{4}{5} =$$

$$\frac{1}{3} \div \frac{5}{8} = \qquad \frac{4}{7} \div \frac{2}{3} = \qquad \frac{2}{5} \div \frac{3}{8} = \qquad \frac{4}{5} \div \frac{1}{2} =$$

$$\frac{3}{7} \div \frac{2}{9} = \qquad \frac{4}{9} \div \frac{5}{9} =$$

It is not always easy to see which is the bigger in a group of fractions. It is important to know which is bigger, however, especially when you are taking away one fraction from another, when you may need to put the bigger fraction first and take the smaller fraction away from it.

To find out which is the bigger fraction you must change them so that they have the same denominator.

For example:

$$\frac{1}{2} \qquad \frac{1}{4} \qquad \text{which is bigger?}$$

Remember that the larger the number at the bottom, the smaller the fraction (if the number at the top is the same).

Change $\frac{1}{2}$ into $\frac{2}{4}$ by multiplying by **2** top and bottom.

You can now see that $\frac{2}{4}$ is bigger than $\frac{1}{4}$

Which is the biggest in these groups of fractions?

1. $\frac{1}{2}$ $\qquad\qquad$ $\frac{3}{4}$ $\qquad\qquad$ $\frac{1}{5}$

2. $\frac{3}{10}$ $\qquad\qquad$ $\frac{2}{3}$ $\qquad\qquad$ $\frac{4}{9}$

3. $\frac{1}{8}$ $\qquad\qquad$ $\frac{1}{4}$ $\qquad\qquad$ $\frac{1}{2}$

Simplifying fractions is a way of making awkward fractions easier to handle. For instance:

$\frac{14}{49}$ can be simplified to $\frac{2}{7}$ by dividing the top number (14) and the bottom number (49) by 7.

You choose the number to divide by but it must be a number which will go exactly into both top and bottom numbers.

In this case **7** will do this, **14 ÷ 7 = 2** and **49 ÷ 7 = 7** so you get the answer $\frac{2}{7}$

Sometimes you'll find the fraction you get after simplifying can be simplified still further as in this example:

$$\frac{36}{72}$$

If you choose to divide by **6** you say **36 ÷ 6 = 6** and **72 ÷ 6 = 12** and get the fraction

$$\frac{6}{12}$$

$\frac{6}{12}$ can be simplified further by saying **6 ÷ 6 = 1** and **12 ÷ 6 = 2**, so the answer is $\frac{1}{2}$

This is as far as you can go.

This is called the lowest form of the fraction – when the only number that will divide into both the top and bottom number is **1**.

Now try simplifying these fractions:

$$\frac{2}{10} \qquad \frac{3}{9} \qquad \frac{5}{15}$$

$$\frac{6}{48} \qquad \frac{4}{24} \qquad \frac{8}{16}$$

$$\frac{5}{20} \qquad \frac{9}{45} \qquad \frac{7}{56}$$

$$\frac{16}{32} \qquad \frac{25}{55} \qquad \frac{77}{99}$$

Addition

When a sum includes a whole number and a fraction, like this:

$$1\tfrac{1}{2} + 2\tfrac{1}{3} =$$

you can simply add the whole numbers, **1** and **2**, making **3**, and then add the fractions in the usual way, like this:

$$3 \quad \frac{3 + 2}{6} = 3\tfrac{5}{6}$$

Take away, multiplication and division

If the sum is take away, multiply or divide, it's safer to turn the fraction and whole number into a top-heavy fraction (called an improper fraction).

Look at this sum: $2\tfrac{1}{2} \times 1\tfrac{2}{3} =$

The first thing to do is to change the $2\tfrac{1}{2}$ and the $1\tfrac{2}{3}$ into improper fractions like this:

You need to find the number of halves in $2\tfrac{1}{2}$

To do that you say $2 \times 2 = 4$ (this gives you the number of halves in **2**).

Then add **1** for the extra $\tfrac{1}{2}$, this makes **5** halves in $2\tfrac{1}{2}$

So: $2 \times 2 = 4$ and $4 + 1 = 5$ and $2\tfrac{1}{2} = \tfrac{5}{2}$

Now change $1\tfrac{2}{3}$ into an improper fraction.

You need to find the number of thirds in $1\tfrac{2}{3}$

To do that you say $3 \times 1 = 3$ (this gives you the number of thirds in **1**).

Then add **2** for the extra $\tfrac{2}{3}$, this makes **5** thirds in $1\tfrac{2}{3}$

So: $3 \times 1 = 3$ and $3 + 2 = 5$ and $1\tfrac{2}{3} = \tfrac{5}{3}$

Now you can do the sum in the usual way: $2\tfrac{1}{2} \times 1\tfrac{2}{3} = \tfrac{5}{2} \times \tfrac{5}{3} = 2\tfrac{25}{6}$

To turn $\tfrac{25}{6}$ into a whole number and fraction you work out $25 \div 6 = 4\tfrac{1}{6}$

So the answer is $4\tfrac{1}{6}$

As the old measurements are phased out, fractions will be used less and less. Here are a few problems where knowledge of fractions is still required.

1. If 1lb of cheese is shared between 4 people, how much will they each have?

2. Mrs Popat has 5 children. She buys 2 bars of chocolate to share between them. Each bar has 6 pieces. What is the fairest way to share out the chocolate?

3. I want to knit a scarf measuring 20cm wide. My tension is $4\frac{1}{2}$ stitches to 5cm. How many stitches must I cast on?

4. My recipe book shows $\frac{1}{2}$ lb of flour, $\frac{1}{4}$ lb fat, and a small amount of salt and water as the ingredients for short crust pastry. I want to make double the quantity – how much flour and fat should I use?

5. Apples on a market stall are priced at 90p a kilo or 40p a pound. Is this correct (a kilo is approximately $2\frac{1}{4}$ lbs).

6. I was a quarter of an hour late for an appointment, and then had to spend half an hour looking for somewhere to park. How late was I in the end?

7. Mr Ali gets time and a half when he works overtime on Saturdays. His usual rate is £5.90 an hour for a 40-hour week. If he works a full 8 hour shift on Saturday, how much will he earn that week?

Decimals and fractions

Decimals are part of the place value system (see Book 1, page 12). A decimal number is any number to the right of the decimal point. Any number in this position is worth less than one.

Decimal is short for decimal fraction. Decimals are just another way of expressing fractions or parts of one, expressed as tenths, hundredths, thousandths, and so on.

The first number to the right of the decimal point is a number of tenths, the second number is a number of hundredths and so on.

For example:

0.1 is the same as $\frac{1}{10}$ and **0.01** is the same as $\frac{1}{100}$

2.45 is $\frac{245}{100}$ or **2** and $\frac{4}{10}$ and $\frac{5}{100}$

Hundredths

0.2	$\frac{20}{100} = \frac{2}{10}$
0.19	$\frac{19}{100} = \frac{1}{10} + \frac{9}{100}$
0.18	$\frac{18}{100} = \frac{1}{10} + \frac{8}{100}$
0.17	$\frac{17}{100} = \frac{1}{10} + \frac{7}{100}$
0.16	$\frac{16}{100} = \frac{1}{10} + \frac{6}{100}$
0.15	$\frac{15}{100} = \frac{1}{10} + \frac{5}{100}$
0.14	$\frac{14}{100} = \frac{1}{10} + \frac{4}{100}$
0.13	$\frac{13}{100} = \frac{1}{10} + \frac{3}{100}$
0.12	$\frac{12}{100} = \frac{1}{10} + \frac{2}{100}$
0.11	$\frac{11}{100} = \frac{1}{10} + \frac{1}{100}$
0.1	$\frac{10}{100} = \frac{1}{10}$
0.09	$\frac{9}{100}$
0.08	$\frac{8}{100}$
0.07	$\frac{7}{100}$
0.06	$\frac{6}{100}$
0.05	$\frac{5}{100}$
0.04	$\frac{4}{100}$
0.03	$\frac{3}{100}$
0.02	$\frac{2}{100}$
0.01	$\frac{1}{100}$
0	

Tenths

1.5	$1\frac{5}{10} = 1\frac{1}{2}$
1.4	$1\frac{4}{10} = 1\frac{2}{5}$
1.3	$1\frac{3}{10}$
1.2	$1\frac{2}{10} = 1\frac{1}{5}$
1.1	$1\frac{1}{10}$
1.0	$\frac{10}{10} = 1$
0.9	$\frac{9}{10}$
0.8	$\frac{8}{10} = \frac{4}{5}$
0.7	$\frac{7}{10}$
0.6	$\frac{6}{10} = \frac{3}{5}$
0.5	$\frac{5}{10} = \frac{1}{2}$
0.4	$\frac{4}{10} = \frac{2}{5}$
0.3	$\frac{3}{10}$
0.2	$\frac{2}{10} = \frac{1}{5}$
0.1	$\frac{1}{10}$
0	

Place value: decimals

Fit the numbers below into the correct place on the place value chart (the first one is done for you).

	Tens	Units	Tenths	Hundredths	
	T	U	$\frac{1}{10}$	$\frac{1}{100}$	
24.25	2	4	2	5	twenty-four point two five
69.16					
8.49					
60.23					
35.49					
1.26					
0.5					
28.16					
94.03					

Which is the bigger decimal?

0.2 or 0.02	0.04 or 5.3
3.14 or 31.4	8.6 or 5.3
0.6 or 0.3	3.79 or 3.709
7.08 or 7.8	2.8 or 2.75
3.6 or 3.64	7.1 or 5.06
0.05 or 0.005	0.06 or 0.032
3.1 or 3.11	17.6 or 17.623
0.2 or 0.04	20.1 or 20.004
0.54 or 0.504	

Write these numbers

1. twenty-eight point one

2. thirty-two point four

3. sixty-nine point two five

4. three hundred and twenty-one point three

5. seventy-nine

6. two point 0 five

7. one hundred and three point 0 two

8. ninety-seven point 0 six

9. two hundred and forty-eight point four

10. eleven point zero seven nine

11. thirty-five point zero zero seven

12. three hundred and forty-five

13. fifty-five point 0 five

14. eighteen point two four six

15. sixteen point nought seven

Place value: decimal chart

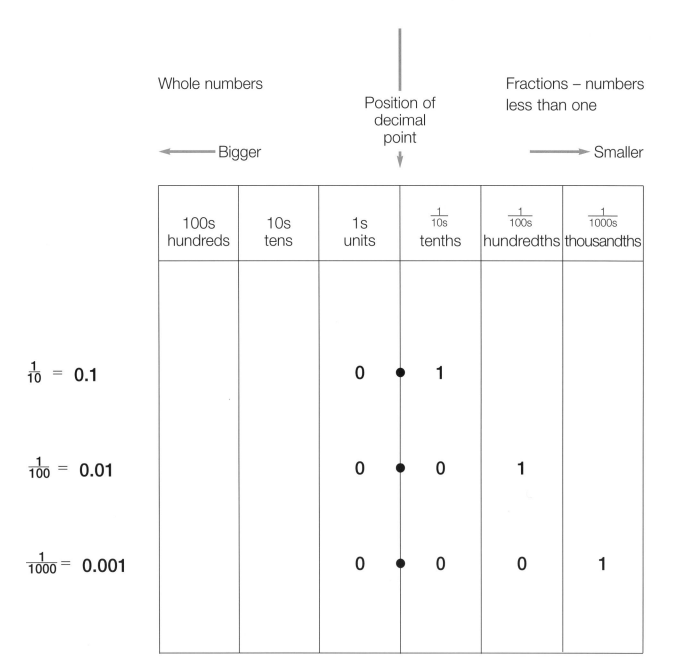

Whole numbers

Position of decimal point

Fractions – numbers less than one

◀——— Bigger

——▶ Smaller

100s hundreds	10s tens	1s units	$\frac{1}{10s}$ tenths	$\frac{1}{100s}$ hundredths	$\frac{1}{1000s}$ thousandths
		0	1		
		0	0	1	
		0	0	0	1

$\frac{1}{10} = 0.1$

$\frac{1}{100} = 0.01$

$\frac{1}{1000} = 0.001$

There is one number in each column to show the value of that number.

Multiplying and dividing decimals by 10, 100 or 1000 is nowhere near as hard as it looks.

You need to remember how the place value system works with whole numbers since decimals work in exactly the same way, and are part of the same system. Look at this example:

$$5 \times 10 = 50$$

The **5** moves one place to the left and is worth ten times what it was worth in the units column.

The same thing happens when the number being multiplied is **0.5**

$$0.5 \times 10 = 5$$

The **5** has moved one place to the left and has become **5** units rather than **5** tenths ($\frac{5}{10}$)

If we multiply by 100 the number or numbers move two places to the left, like this:

$$2 \times 100 = 200$$

and

$$0.2 \times 100 = 20$$

When dividing, the answer will be smaller, so you move the numbers to the right instead of to the left, so:

$$40 \div 10 = 4$$

and

$$4 \div 10 = 0.4$$

$$0.4 \div 10 = 0.04$$

$$0.04 \div 100 = 0.0004$$

Try these multiplication and division sums:

$2.5 \times 10 =$	$5.9 \times 100 =$	$6.3 \times 10 =$
$1.8 \times 10 =$	$3.5 \div 10 =$	$7.8 \div 100 =$
$27.4 \times 100 =$	$50.3 \times 10 =$	$45.7 \div 10 =$
$190.2 \times 10 =$	$70.1 \div 10 =$	$640.5 \times 10 =$

× 10, 100, 1000

When you multiply by 10 just move the figures one place to the left.

When you multiply by 100 just move the figures two places to the left.

When you multiply by 1000 just move the figures three places to the left.

Remember, the figures may have to jump over the decimal point.

1. $10 \times 10 =$

2. $69 \times 10 =$

3. $148 \times 100 =$

4. $32 \times 10 =$

5. $51 \times 1000 =$

6. $78.5 \times 100 =$

7. $20.25 \times 10 =$

8. $72.1 \times 100 =$

9. $4.38 \times 100 =$

10. $16.5 \times 100 =$

11. $5.94 \times 10 =$

12. $17.1 \times 10 =$

13. $0.06 \times 100 =$

14. $3.002 \times 100 =$

15. $2.15 \times 1000 =$

16. $48.7 \times 100 =$

17. $12.6 \times 10 =$

18. $192.5 \times 100 =$

19. $36.4 \times 1000 =$

20. $4.29 \times 10 =$

÷ 10, 100, 1000

When you divide by 10 just move the figures one place to the right.

When you divide by 100 just move the figures two places to the right.

When you divide by 1000 just move the figures three places to the right.

Remember, the figures may have to jump over the decimal point.

1. $160 \div 10 =$

2. $200 \div 100 =$

3. $40 \div 1000 =$

4. $390 \div 10 =$

5. $45 \div 10 =$

6. $18 \div 10 =$

7. $98 \div 100 =$

8. $28.6 \div 10 =$

9. $35.5 \div 100 =$

10. $11.2 \div 1000 =$

11. $120 \div 10 =$

12. $17.5 \div 100 =$

13. $149.6 \div 10 =$

14. $528.2 \div 10 =$

15. $39.25 \div 1000 =$

16. $14.86 \div 10 =$

17. $439.2 \div 1000 =$

18. $205.6 \div 100 =$

19. $78.06 \div 10 =$

20. $13.05 \div 1000 =$

Write these in words – fractions and decimals

$$\frac{3}{4} + \frac{3}{8} = 1\frac{1}{8}$$

$$3.6 = 3\frac{6}{10}$$

$$5.4 \div 9 = 0.6$$

$$2\frac{6}{7} - \frac{5}{7} = 2\frac{1}{7}$$

$$3\frac{5}{6} = \frac{23}{6}$$

$$\frac{1}{10} + \frac{2}{100} = 0.12$$

Write these in numbers – fractions and decimals

Write these sentences using numbers and signs instead of words.

1. Five point three is the same as five and three tenths.

2. Three point three four plus nought point two one makes three point five five.

3. Four and four-fifths minus nine tenths leaves three and nine tenths.

4. Twenty-five quarters equals six and a quarter.

5. Two thirds times three-eighths equals six twenty-fourths.

6. Twenty-two point three six minus nine point seven leaves twelve point six six.

7. One third divided by five eighths is the same as one third times eight fifths.

8. Thirty-two hundredths equals nought point three two.

How to add decimals

Just remember to keep the decimal points lined up underneath each other.

For example: 0.5 + 0.25

Set it up like this: 0.50 +
                                          0.25

You can always put an extra 0 in a gap to keep the columns in line.

Then you just do it like an ordinary adding-up sum.

                                          0.50 +
                                          0.25
                                          0.75

The decimal point goes underneath the other decimal points and the answer is **0.75**

You say: nought point seven-five, never nought point seventy-five, because that makes it sound like a whole number.

Now try these:

0.25 +   0.75 +   0.25 +
0.25   0.50   0.75

0.80 +   0.35 +   0.78 +
0.91   0.25   0.36

As with adding-up decimals the important thing is to keep the decimal points lined up underneath each other.

The only thing you have to be careful of is that, as in any ordinary take-away sum, you must have the bigger number on top.

For example: $\qquad$ **0.75 − 0.25**

Look at the number on the right next to the decimal point, **7** is a bigger number than **2**, so set it out like this:

$$0.75 -$$
$$\underline{0.25}$$

Then do it like an ordinary take-away sum.

$$0.75 -$$
$$\underline{0.25}$$
$$0.50$$

Put the decimal point in your answer under the other decimal points.

The answer is **0.50** or just **0.5**.

You say: nought point five.

Now try these:

0.50 − 0.25	0.75 − 0.50	0.25 − 0.10
0.78 − 0.35	0.92 − 0.50	0.63 − 0.47

Decimals – add and take away

Add:

$$5.01 + \\ 3.56$$

$$0.80 + \\ 6.07 \\ 0.90$$

$$2.312 + \\ 0.041 \\ 1.206$$

Set down in columns and add:

$0.48 + 6.9 + 0.07$

$2.06 + 3.5 + 4.44$

$12.75 + 0.9 + 8.67$

$0.6007 + 0.089 + 2.5$

Take away:

$$5.06 - \\ 4.07$$

$$3.30 - \\ 1.05$$

$$2.280 - \\ 0.028$$

$$7.30 - \\ 6.91$$

Take away the smaller number from the larger:

7 and **4.6**

3.02 and **0.801**

0.75 and **0.9**

0.093 and **0.41**

0.6 and **0.37**

1.03 and **0.993**

2.01 and **0.83**

How to multiply decimals

To multiply a decimal number by a whole number

Look at these sums:

12 ×	12 ×	12 ×
3	0.3	0.03
36	3.6	0.36
whole numbers	one decimal place	two decimal places

The only difference is the position of the decimal point.

Count the number of figures after the point, and put the point in the same place in the answer (counting from the right).

Try these: **15 × 0.5**   **24 × 9.8**   **152 × 2.7**   **260 × 1.23**

To multiply two decimal numbers

3.4 × 0.6 Make a guess at the result so you can check your answer. To do this, round the numbers off to the nearest whole ones, here 3 × 1 = 3

To work out exactly: first ignore the decimal points, and work the sum in the usual way. Then put back the point in the answer by counting the number of decimal places in both numbers.

3.4 ×	34 ×
0.6	6
2.04	204

Here there are two numbers after the point, so there must be the same in the answer, counting from the right. The answer is 2.04.

Always check your answer against the guess you made at the start.

Let's try: **31.23 × 1.7**   Rough guess: **31 × 2 = 62**

Work it out this way:

3123 ×		31.23 ×	3 decimal places
17		1.7	
31230		53.091	3 decimal places
21861			
53091			

The exact answer is 53.091 because there are 3 figures after the point in the sum, so there must be 3 figures after the point in the answer.

Now try these: **15.1 × 2.3**   **84.6 × 0.7**   **124.2 × 3.06**

25.63 × 1.3   **181.64 × 0.03**   **16.02 × 3.11**

N2/L1.5

To divide a decimal by a whole number, as in this example, estimate your answer before you start.

$$9 \overline{)27.09}$$

In this case $27 \div 9 = 3$ so you know the answer will be 3 whole ones and a decimal.

Now just work through the sum in the usual way, taking care to keep the decimal point in the answer lined up with the decimal point in the number you are dividing (27.09). You should get the answer 3.01.

To divide a decimal by a decimal, as in this example, again, estimate first, using whole numbers.

$$3.2 \overline{)6.4}$$

$6 \div 3 = 2$, so you know before you start the answer will be 2, possibly with a decimal.

To work out the exact answer you must change the numbers you are dividing by into a whole number to make it easier to handle, using your knowledge of place value. To do this you look at the sum and see how many decimal places there are in the number you are dividing by (3.2).

In this case there is one decimal place (.2) so you must multiply by 10. This will give you a whole number (32) to divide by, rather than a whole number and a decimal. If the number you were dividing by had had two decimal places in it instead of one, then you would have multiplied by 100, and so on.

Now multiply the number you are dividing (6.4) by 10. It's most important that you multiply both the 3.2 and the 6.4 by the same number. As long as you do the same thing to both parts of the sum the relationship between them remains the same.

Now re-write the sum:

$$32 \overline{)64}$$

Now just work it out in the usual way as a long division sum. The answer is 2.
This may seem strange but try it and see.

Now try these:

$$4.2 \overline{)8.4} \qquad 2.5 \overline{)12.5} \qquad 1.2 \overline{)0.8412}$$

3.6 ×
1.4
_____

18.3 ×
0.6
_____

31.5 ×
0.02
_____

16.37 ×
2.4
_____

19.32 ×
4.4
_____

8.007 ×
3.4
_____

92.6 ×
3.25
_____

7.063 ×
2.3
_____

$4)\overline{0.216}$

$7)\overline{0.504}$

$9)\overline{3.87}$

$0.4)\overline{4.8}$

$0.5)\overline{3.05}$

$1.1)\overline{25.3}$

$0.6)\overline{0.084}$

$2.1)\overline{44.1}$

$1.8)\overline{57.6}$

N2/L1.5, N2/L2.6

Rounding off, also called simplifying or approximating, is a way of making awkward numbers easier to handle. You decide how accurate you need to be in each case when rounding off.

For example, when shopping in the supermarket you don't want to find you've got more goods in your basket than you can afford to pay for. You could add up the prices with a calculator as you go round the shop, but if you are adding things up in your head it's a lot easier if you round all the prices up to the next 10p, so 65p becomes 70p, £5.90 becomes £6.00 and so on. If you do this you can be sure you won't overspend as you've rounded up to the nearest ten pence or pound.

However, if you're estimating the size of the crowd at a football match you need to be accurate to the nearest thousand, rather than the nearest ten people.

We call the rounded off numbers 'round numbers', so in the examples given 70p and £6.00 are round numbers. In the case of the football crowd 40,000 would be a reasonable round figure.

Rounding off with decimals

Rounding off is particularly useful when dealing with decimals.

For example: **3.9** becomes **4** correct to the nearest whole number.

2.68 becomes **2.7** correct to one decimal place. We could round off still further and make it **3** correct to the nearest whole number.

The rule is that if the figure at the far right end of the number is 5, 6, 7, 8 or 9 then the figure next to it on the left is rounded up by one.

If the figure on the far right end of the number is 0, 1, 2, 3 or 4, then the number next to it on the left stays as it is.

You can do this as many times as you need to, in order to be as accurate as necessary. Remember that the more times you 'round off' in this way the less accurate your answer becomes. This is why it is important to know how accurate you need to be in each case.

If you're working through a calculation involving several stages and you need a fairly accurate answer to your calculation, don't round off until the end.

Try rounding off these numbers correct to one decimal place:

6.88	5.16	8.92
9.134	10.026	112.55

Decimal problems

1. How much do 6 litres of petrol cost at 84p per litre?

2. I want to check my electricity bill, which is charged on 3 tariffs, low, medium and high. I have used 252 high-tariff units at 9.54 pence per unit. I have also used 309 units at 6.76 pence each and 235 low-tariff units at 2.77 pence. My bill was £51.44. Is this correct?

3. The telephone bill for calls on my landline is £70.73 and the service charge is £39.41. VAT at 17.5% is added to the total, giving an overall total of £129.53. Is this correct?

4. Gas bills are calculated on 2 rates, Band A and Band B. The unit is called a kilowatt hour (kwh). Band A units are 2.110 pence and Band B units are 1.143 pence. I've used 1718 Band A units and 1157 Band B units. The total gas charge on my bill is £49.47. How much of that is Band A and how much is Band B?

5. I want to change £120.00 into US dollars (USD or US $). If the exchange rate is $1.83 to the pound sterling (£ or GBP), how much should I get in dollars?

6. To convert from miles to kilometres, you multiply by 1.6093. How many kilometres are there in $12\frac{1}{2}$ miles?

Per cent

Per cent means 'in each hundred'.

Per cent is written like this %.

100% means 100 in each 100, or the whole of something.

50% means 50 in each 100, or half of something.

25% means 25 in each 100, or a quarter of something.

75% means 75 in each 100, or three quarters of something.

Try these:

100% of £100 is ....

50% of £10 is ....

25% of £400 is ....

75% of £100 is ....

100% of 900 people is ....

50% of 400 people is ....

25% of 800 people is ....

Percentages and fractions

Per cent means 'in each hundred'.

50% means 50 out of every hundred.

As a fraction this is written $\frac{50}{100}$ which is the same as $\frac{1}{2}$

In the same way:

$$1\% = \frac{1}{100}$$

$$100\% = \frac{100}{100} = 1 \text{ (the whole of something)}$$

$$25\% = \frac{25}{100} = \frac{1}{4}$$

Percentages to fractions

Any percentage can be written as a fraction – write the number as a fraction over 100, and simplify if possible.

$$17\% = \frac{17}{100} \qquad\qquad 15\% = \frac{15}{100} = \frac{3}{20} \qquad\qquad 22\% = \frac{22}{100} = \frac{11}{50}$$

$$12\frac{1}{2}\% = \frac{12\frac{1}{2}}{100} = \frac{25}{200} = \frac{1}{8} \qquad\qquad 7\frac{1}{2}\% = \frac{7\frac{1}{2}}{100} = \frac{15}{200} = \frac{3}{40}$$

Try these:

13%	30%	5%	10%	$2\frac{1}{2}\%$	33%
75%	68%	$15\frac{1}{2}\%$	$6\frac{1}{4}\%$	9%	66%

Fractions to percentages

Any fraction can be written as a percentage – multiply the fraction by 100 ($\frac{100}{1}$), and call it the percentage.

$$\frac{1}{2} \qquad\qquad \frac{1}{2} \times \frac{100}{1} = \frac{100}{2} = 50 \qquad\qquad \text{so } \frac{1}{2} = 50\%$$

$$\frac{1}{100} \qquad\qquad \frac{1}{100} \times \frac{100}{1} = \frac{100}{100} = 1 \qquad\qquad \text{so } \frac{1}{100} = 1\%$$

$$\frac{1}{4} \qquad\qquad \frac{1}{4} \times \frac{100}{1} = \frac{100}{4} = 25 \qquad\qquad \text{so } \frac{1}{4} = 25\%$$

$$\frac{7}{8} \qquad\qquad \frac{7}{8} \times \frac{100}{1} = \frac{700}{8} = 87\frac{4}{8} = 87\frac{1}{2} \qquad\qquad \text{so } \frac{7}{8} = 87\frac{1}{2}\%$$

Try these:

$\frac{3}{4}$	$\frac{1}{10}$	$\frac{1}{5}$	$\frac{23}{100}$	$\frac{3}{5}$	$\frac{1}{20}$	$\frac{5}{100}$	$\frac{1}{8}$
$\frac{5}{8}$	$\frac{91}{100}$	$\frac{17}{40}$	$\frac{2}{3}$	$\frac{1}{6}$	$\frac{5}{12}$	$\frac{3}{200}$	

To calculate percentages using this method you need to understand that 100% means the whole of something, and work from there.

If you want to find: **50%** of **600** people, you say:

100% of **600** people = **600 people**

50% of **600** people = **300 people**

You've done a hidden division sum.

You've said that **100% ÷ 2 = 50%** and done the same thing at the other end of the equation: **600 people ÷ 2 = 300 people**.

You divide by **2** in this case because you know that you have to divide **100** by **2** to get the answer **50**.

Suppose you want to find: **25%** of **600** people, you say:

100% of **600** people = **600 people**,

25% of **600** people = **150 people**

You've divided by 4 this time because 100 ÷ 4 = 25, so you divide 600 by 4 to get 150.

Now try these:

50% of £40	25% of 800
75% of 1000	10% of £500
75% of 200 people	50% of 1000 votes
25% of 80 countries	15% of 1000 cars
40% of 160 marks	

VAT

You can use the practice method to work out VAT at $17\frac{1}{2}$% if you haven't got a calculator to hand.

To find $17\frac{1}{2}$% of a sum of money, first divide by **10** to find **10%**, then halve it to find **5%**, then halve that again to find $2\frac{1}{2}$%. Add the three figures together and you've got $17\frac{1}{2}$%, which you can then add to the nett figure (the cost without VAT).

N2/L1.8

Finding percentages – fraction method

What is **20%** of **635**?

How do I find **12%** of **£6.50**?

How much is $7\frac{1}{2}$ **%** interest on **£1,250**?

First change the percentage into a fraction, replace 'of' by $\times$ and multiply.

20% of **635** $\qquad \frac{20}{100} \times 635 \qquad \frac{1}{5} \times \frac{635}{1} = \frac{635}{5} = 127 \qquad$ so **20%** of **635** is **127**

12% of **£6.50**

First change **£6.50** to **650p**

$$\frac{12}{100} \times 650 = \frac{780}{10} = 78 \qquad \text{so } 12\% \text{ of } £6.50 \text{ is } 78p$$

$7\frac{1}{2}$ **%** of **£1,250.** We can leave this sum of money in pounds.

$$\frac{7\frac{1}{2}}{100} \times 1250 = \frac{15}{200} \times 1250 = \frac{3}{4} \times 125 = \frac{375}{4} = 93\frac{3}{4}$$

so $7\frac{1}{2}$% of **£1250** is **£93** $\frac{3}{4}$ which is **£93.75**

Try these:

5% of 900 $\qquad\qquad\qquad$ 12% of £38.00 $\qquad\qquad\qquad$ 32% of 525

$2\frac{1}{2}$ % of £16.20 $\qquad\qquad\qquad$ $7\frac{1}{2}$ % discount on £2,320

What percentage is **15** out of **300**?

First write the numbers as a fraction, then multiply by 100, and the answer is the percentage.

$$\frac{15}{300} \times \frac{100}{1} = \frac{1500}{300} = \frac{15}{3} = 5\%$$

Finding percentages – fraction method – continued

What percentage is **5p** out of **£5**?

Before we can work the percentage, both numbers must be in the **same units**, so here we change £s to pence: **£5 = 500p**

$$\frac{5}{500} \times \frac{100}{1} = \frac{500}{500} = \frac{1}{1} = 1\%$$

Find these percentages:

65 out of 650   18 out of 30   50 out of 250

15p in the pound   £1.50 discount on £75.00

N2/L1.8, N2/L1.9

Finding percentages – using a calculator

Many calculators have a percentage button, marked **%**.

For example, to find **50%** of **40**, key in **40 × 50** and then press **%**. The percentage button automatically divides by 100, so the answer will show as **20** (i.e. **20%**).

Alternatively, you can press **40 × 50** = which will give you **2000**, and then divide by **100**, which will also give you the correct answer: **20**.

You could also say **40** times **0.5**. The answer will also be **20**.

This is because **0.5** is equivalent to **50** divided by **100**, or a half (see 'Comparing percentages, fractions and decimals' on p44 in this book).

N2/L1.11

Percentage increase and decrease

Percentage increase and decrease are frequently used, for example in the government's budget, in pay negotiations, and in shops with sales promotions.

For example, suppose a bus fare has gone up from **80p** to **£1**

The actual increase is **20p**

What **percentage increase** is this?

The formula is:

$$\% \text{ increase} = \frac{\text{actual increase}}{\text{original number}} \times \frac{100}{1}$$

In this case:

$$\% \textbf{ increase} = \frac{20}{80} \times \frac{100}{1} = \frac{1}{4} \times \frac{100}{1} = \frac{100}{4} = 25\%$$

So there was a **25%** increase in bus fares. The new figure is **125%** of the original price.

When you are doing a sum like this, make sure that the actual increase and the original number are in the **same units**, in this case pence.

Percentage decrease is worked out in the same way.

A digital camera is reduced from **£120** to **£84** in a sale. What percentage decrease is this?

The formula is:

$$\% \text{ decrease} = \frac{\text{actual decrease}}{\text{original number}} \times \frac{100}{1}$$

In this case, the actual decrease is **£120 − £84 = £36**.

$$\% \textbf{ decrease} = \frac{36}{120} \times \frac{100}{1} = \frac{36}{6} \times \frac{5}{1} = \frac{6 \times 5}{1} = 30\%$$

The camera was reduced by **30%** so it could have been marked **30%** off.

If a store has a promotion saying **30% off all marked prices** you can either find 30% of the original price and then subtract, or you can work out 70% of the original price.

Percentage problems

1. My shoes cost £45 in a sale. The ticket said '50% off', so what was the original price?

2. Alun got a 10% pay rise last week. He used to be paid £120.00 a week basic, what does he get now?

3. Shukri's rent went up 8% last June. She used to pay £140.00 a week for her flat, how much is it now?

4. If you read that the pound (£) was now 60% of its 1981 value, what would that mean exactly?

5. You sometimes hear people say 'I'm 99% sure of it'. What do they mean?

6. How do you say 'double it' as a percentage?

7. You have to put a 10% deposit on a new washing machine as part of the hire purchase agreement. If the machine costs £280, how much is the deposit?

8. In an opinion poll, 5% of those questioned were 'Don't knows'. 1300 people were interviewed altogether. What's 5% of 1300?

9. My son got 40/50 in his maths exam. His teacher said that was 70%. Was he right?

10. My partner got a 3% pay rise. He earns £18,000 per year. I've just got a 2.5% rise on my income of £22,000 per year. Which of us gets the biggest increase?

Problems on percentage increase and decrease

1. A jacket is marked £63, reduced from £75. What percentage decrease is this?

2. My pay rise came through this week. I received £125.50. Last week I got £120. What percentage pay rise have I had?

3. In the budget there's been a 3% increase in fuel tax. It used to cost me £32 to fill my tank, how much will it cost me now?

4. In the sales, a table is marked '15% off'. The original price was £145.99. What is the new price?

5. In a discount store a sign says '20% off all marked prices'. How much would I save on these things?
   a) A set of pans marked £28.95
   b) An electric heater marked £29.20
   c) A blanket marked £28.42

6. I am due for a 9% pay increase. My last pay packet was £187.50. What will I get this week?

7. I saw the same radio for sale in 2 shops. One was marked down from £19.15 to £15.50. The other was marked '12% reduction' from £18.25.
   a) Which was the cheaper radio?
   b) Which was reduced the most?

Comparing percentages, fractions and decimals

Percentages, fractions and decimals are different ways of expressing the same thing – in other words, they are equivalent. For example, 50% is the same as $\frac{1}{2}$ and 0.5.

Percentage	Fraction	Decimal	
100%	1	1.0	
75%	$\frac{3}{4}$	0.75	
50%	$\frac{1}{2}$	0.5	
25%	$\frac{1}{4}$	0.25	
$12\frac{1}{2}$%	$\frac{1}{8}$	0.125	
$33\frac{1}{3}$%	$\frac{1}{3}$	$0.\overset{\bullet}{3}$ (you say nought point three recurring)	
$66\frac{2}{3}$%	$\frac{2}{3}$	$0.\overset{\bullet}{6}$ (you say nought point six recurring)	
10%	$\frac{1}{10}$	0.1	

Percentage	Fraction	Decimal	
20%	$\frac{2}{10}$ or $\frac{1}{5}$	0.2	
30%	$\frac{3}{10}$	0.3	
40%	$\frac{4}{10} = \frac{2}{5}$	0.4	
60%	$\frac{6}{10} = \frac{3}{5}$	0.6	
70%	$\frac{7}{10}$	0.7	
80%	$\frac{8}{10} = \frac{4}{5}$	0.8	
90%	$\frac{9}{10}$	0.9	
1%	$\frac{1}{100}$	0.01	

N2/L1.3

Since fractions, decimals and percentages are all ways of expressing the same thing, it is useful to be able to switch between them.

Changing fractions into decimals

If you want to change a fraction into a decimal, start by looking at the fraction. Say it's $\frac{3}{4}$ that means 3 divided by 4, so you can do it as a division calculation, showing the remainder as a decimal (see 'What do you do if there is a remainder?' in Book 2).

$3 \div 4 = 0.75$

If you do the division on a calculator it will give you the answer as a decimal.

This may be obvious in this case, but try it with $\frac{24}{96}$

That means 24 divided by 96:

$24 \div 96 = 0.25$ so $\frac{24}{96} = 0.25$

Changing fractions into percentages

To change a fraction (say $\frac{3}{4}$) into a percentage, multiply it by 100.

100 is the same as $\frac{100}{1}$ so you can set the calculation out like this:

$$\frac{3}{4} \times \frac{100}{1} = \frac{300}{4} = 75$$

so the answer is **75%**.

This is the sort of calculation you might have to do if you scored 24 out of 96 in a test and wanted to turn the answer into a percentage. In that case, the fraction you'd start with would be $\frac{24}{96}$.

$$\frac{24}{96} \times \frac{100}{1} = \frac{2400}{96} = \frac{25}{1} = 25\%$$

Changing decimals into fractions

If you start with a decimal, say 0.75, and you want to turn it into a fraction, just write the decimal numbers over 100, like this:

$\frac{75}{100}$ and then simplify if necessary.

In this case $\frac{75}{100}$ simplifies down to $\frac{3}{4}$

so **$0.75 = \frac{3}{4}$** .

If you want to know why this works, look at the sections on 'Decimals and fractions' and 'Place value: decimals' in this book.

Changing decimals into percentages

If you start with a decimal, say 0.75, and you want to turn it into a percentage, just multiply it by 100.

$$0.75 \times 100 = 75$$

The answer is a percentage,

so $0.75 = 75\%$.

Changing percentages into fractions

If you want to change a percentage into a fraction, just write the percentage number over 100.

For example, 75% can be written $\frac{75}{100}$ which simplifies down to $\frac{3}{4}$

so $75\% = \frac{3}{4}$

Changing percentages into decimals

If you start with a percentage, say 75%, and you want to turn it into a decimal, just divide it by 100.

$$75 \div 100 = 0.75$$

If you do this on a calculator, the answer will come out as a decimal.

Fractions, decimals and percentages equivalents

$$\frac{75}{100} = 0.75 = 75\%$$

Work out these fraction, decimal and percentage equivalents:

45%	$\frac{20}{400}$	0.67	266%	$\frac{3}{5}$	0.39

N2/L2.2

Ratio and proportion

Ratio expresses the proportional relationship between two or more numbers. The actual numbers, called the terms of the ratio, may change but so long as the relationship between them remains the same, so does the ratio.

Examples

Ratios are used in many contexts. For example, a simple salad dressing has 2 parts of oil to 1 of vinegar. In other words the proportion of oil is twice that of vinegar. This is expressed as a ratio of **2 to 1**, written as **2:1**. You can make any quantity of dressing, so long as you keep the proportions of oil and vinegar the same.

Sand and cement may be mixed in the ratio of **5 to 2** when making cement for building. This is expressed as **5:2**.

A map of a city may be on a scale of **1:20,000**. This means that 1 centimetre (cm) on the map represents 20,000 cm on the ground, but it's more useful to think of every 5 cm on the map representing 100,000 cm, which is equal to 1 kilometre (km) on the ground.

This is because for practical purposes when travelling we measure distances on the ground in kilometres rather than centimetres.

In order to do this, you need to work out how many times 20,000 cm goes into 1 km or 100,000 cm. Therefore 1:20,000 is the same as 5:100,000. The relationship between the two quantities is the same in each case.

A fashion shop may sell 2 medium size tops for every 1 small size. In that case, the ratio of small to medium size tops sold would be expressed as 1:2. The ratio of small, medium, large and extra large trousers sold could be 1:3:2:1. Buyers try to order goods in the same ratio they think their customers will need, so as not to be left with unsold stock.

A 1 litre bottle of orange squash can be diluted to make approximately 7 litres of drink for a children's party. This means that one part of squash is mixed with six parts of water. In other words, it is diluted in a ratio of 1:6 (one to six).

Confusingly, the same dilution may be expressed as 1 in 7, but why?

Think of 1:6 and add the terms of the ratio together: 1 + 6 = 7. In other words, once you've made up the drink there are 7 parts altogether, 6 of which are water and 1 of which is squash. There is 1 part of squash in every 7 parts of the total solution.

So 1:6 means the same as 1 in 7, or one part in seven.

Simplifying ratios

It's a good idea to simplify ratios where possible to make the relationship between the numbers easier to understand and to work with. So long as you divide all parts of the ratio by the same number you will preserve the relationship between the quantities. The principle is the same as when simplifying fractions and for the same reason – you're maintaining the relationship between the quantities.

For example, suppose a primary school has 10 teachers on the staff and then 2 leave. At the same time the number of pupils goes down from 250 to 240. The previous ratio of teachers to pupils was 10:250, the new ratio will be 8:240.

If you simplify these ratios using the technique of dividing each quantity in the ratio by the same number, you find that it has gone up from 1:25 to 1:30, i.e., there are 5 more pupils to every teacher. Simplifying the ratios in this way makes it easier to compare the two situations and understand what's going on.

Separating a quantity according to a given ratio

Suppose you are responsible for ordering items for a company operating on 2 sites. You need to buy enough for both sites and then divide the order so as to reflect the scale of the operation in each one. You order 3000 items. The turnover of the larger site is twice that of the smaller one, so you split the order in a ratio of 2:1.

Add the terms of the ratio together to find the total number of parts, in this case $2 + 1 = 3$

Find what fractional part each term is of the whole, in this case $\frac{2}{3}$ and $\frac{1}{3}$

Divide the total quantity (3000) into parts corresponding to the fractional parts. In this case, two thirds of 1000.

$$\frac{2}{3} \times 3000 = 2000 \text{ and } \frac{2}{3} \times 3000 = 1000$$

Ratios, division, percentages, fractions and decimals

A ratio is another way of writing a division sum or a fraction. On some calculators and in some countries, the division sign (÷ or /) is shown as:

1:8 could be written **1 ÷ 8, 1/8** or $\frac{1}{8}$

Just like other fractions, ratios can be expressed as percentages or decimals.

Ratios may also be expressed by the word 'per', as in miles per hour (mph) and revolutions per minute (rpm).

N1/L1.7, N1/L2.3

1. If you want to make 21 kilos of mixed cement using a 5:2 mixture, how much sand and cement do you need to start with?

2. I'm planning a fitted kitchen using a scale drawing at a ratio of 1:40, and I need to accommodate my existing cooker, which is 600mm wide. How can I check whether my cooker will fit into the kitchen?

3. I'm going to feed my plants with a concentrated plant food. The label says mix one part plant food with ten parts water. My watering can holds 4.4 litres, how much plant food should I add to each can?

4. I put a bet on a horse at 5:2. The horse won. My stake was £2.50. How much did I win?

5. What is the staff:student ratio in a college with 3000 students and 120 teaching staff?

6. A shop bought 14 dozen T-shirts in 2 colours, red and white, sizes small, medium, large and extra-large 1:3:2:1. They bought twice as many white T-shirts as red ones. How many T-shirts in each colour and size did they buy?

7. There are 400 people working in a factory. In total, 320 work on 2 shifts during the day and the rest work the night shift. What is the ratio of day to night shift workers?

Fraction block 1 $\frac{1}{2}$ $\frac{1}{4}$ $\frac{1}{8}$, p.7

$\frac{1}{2}$ is half of one.
$\frac{1}{4}$ is half of a half.
$\frac{1}{8}$ is half of a quarter.
$\frac{1}{8}$ is a quarter of a half.
There are 8 eighths in one.
There are 4 eighths in a half.
There are 4 quarters in one.
There are 2 quarters in a half.

Fraction block 1 $\frac{1}{3}$ $\frac{1}{6}$ $\frac{1}{12}$, p.8

A third of one is $\frac{1}{3}$.
$\frac{1}{6}$ is a half of a third.
$\frac{1}{12}$ is a half of a sixth.
There are 6 sixths in one.
There are 2 sixths in a third.
There are 4 twelfths in a third.
There are 8 twelfths in two thirds.
There are 12 twelfths in one.

Fraction block 1 $\frac{1}{5}$ $\frac{1}{10}$, p.9

$\frac{1}{5}$ is a fifth of one.
$\frac{1}{10}$ is a half of a fifth.
A whole one is ten tenths.
There are 5 fifths in one.
There are 2 tenths in a fifth.
There are 6 tenths in three fifths.
There are 10 tenths in one.

Changing fractions $\frac{1}{2}$ $\frac{1}{4}$ $\frac{1}{8}$ $\frac{1}{16}$, p.10

$\frac{1}{4} = \frac{2}{8} = \frac{4}{16}$

$\frac{3}{4} = \frac{6}{8} = \frac{12}{16}$

$\frac{1}{2} = \frac{2}{4} = \frac{4}{8} = \frac{8}{16}$

Changing fractions: $\frac{1}{3}$ $\frac{1}{6}$ $\frac{1}{12}$ and $\frac{1}{5}$ $\frac{1}{10}$, p.11

$\frac{1}{6} = \frac{2}{12}$ $\qquad$ $\frac{5}{6} = \frac{10}{12}$ $\qquad$ $\frac{2}{3} = \frac{4}{6} = \frac{8}{12}$

$\frac{2}{5} = \frac{4}{10}$ $\qquad$ $\frac{3}{5} = \frac{6}{10}$ $\qquad$ $\frac{8}{10} = \frac{4}{5}$ $\qquad$ $\frac{7}{5} = \frac{14}{10}$

How well can you work with fractions?, p.14

1. $\frac{1}{4} \div 5 = \frac{1}{20}$

2. $\frac{1}{2} + \frac{2}{3} - \frac{3}{4} = \frac{5}{12}$

3. $\frac{1}{2} \times \frac{1}{3} = \frac{1}{6}$

4. $\frac{5}{8} \div \frac{2}{3} = \frac{15}{16}$

5. $3 - 1\frac{1}{9} = \frac{17}{9} = 1\frac{8}{9}$

6. $\frac{2}{3} \times 2 = \frac{4}{3} = 1\frac{1}{3}$

7. $1 - \frac{1}{4} = \frac{3}{4}$

8. $3\frac{1}{3} + \frac{2}{3} = 4$

9. $2 + \frac{2}{5} = 2\frac{2}{5}$

10. $1\frac{1}{2} \div 3\frac{3}{4} = \frac{2}{5}$

11. $\frac{15}{16} \times \frac{6}{10} = \frac{9}{16}$

12. $1 + \frac{1}{2} = 1\frac{1}{2}$

13. $\frac{1}{6} \times 5 = \frac{5}{6}$

14. $2\frac{1}{3} + 1\frac{5}{6} + 1\frac{1}{3} = 5\frac{3}{6} = 5\frac{1}{2}$

How to add fractions, p.15

$\frac{1}{4} + \frac{1}{4} = \frac{2}{4} = \frac{1}{2}$ $\qquad$ $\frac{2}{7} + \frac{4}{7} = \frac{6}{7}$ $\qquad$ $\frac{3}{10} + \frac{7}{10} = \frac{10}{10} = 1$

$\frac{1}{9} + \frac{1}{9} = \frac{2}{9}$ $\qquad$ $\frac{2}{3} + \frac{1}{6} = \frac{5}{6}$ $\qquad$ $\frac{2}{3} + \frac{1}{2} = 1\frac{1}{6}$

$\frac{3}{4} + \frac{5}{8} = 1\frac{3}{8}$ $\qquad$ $\frac{1}{5} + \frac{2}{3} = \frac{13}{15}$ $\qquad$ $\frac{5}{6} + \frac{1}{9} = \frac{17}{18}$

$1\frac{1}{4} + 3\frac{1}{3} = 4\frac{7}{12}$ $\qquad$ $2\frac{2}{5} + 4\frac{3}{10} = 6\frac{7}{10}$

$1\frac{3}{8} + 2\frac{1}{4} = 3\frac{5}{8}$ $\qquad$ $6\frac{2}{3} + 2\frac{1}{2} = 9\frac{1}{6}$

How to take away fractions, p.16

$\frac{3}{4} - \frac{1}{4} = \frac{1}{2}$ $\qquad$ $\frac{5}{8} - \frac{3}{8} = \frac{1}{4}$ $\qquad$ $\frac{6}{7} - \frac{1}{7} = \frac{5}{7}$

$\frac{2}{3} - \frac{1}{3} = \frac{1}{3}$ $\qquad$ $\frac{9}{10} - \frac{3}{10} = \frac{6}{10} = \frac{3}{5}$

$\frac{5}{6} - \frac{1}{3} = \frac{1}{2}$ $\qquad$ $\frac{2}{3} - \frac{1}{2} = \frac{1}{6}$ $\qquad$ $\frac{3}{4} - \frac{5}{8} = \frac{1}{8}$

$\frac{3}{5} - \frac{1}{3} = \frac{4}{15}$ $\qquad$ $\frac{3}{4} - \frac{2}{5} = \frac{7}{20}$

$2\frac{1}{4} - 1\frac{3}{4} = \frac{1}{2}$ $\qquad$ $4\frac{1}{8} - 2\frac{1}{4} = 1\frac{7}{8}$ $\qquad$ $3\frac{1}{7} - 1\frac{3}{14} = 1\frac{13}{14}$

$6\frac{3}{10} - 2\frac{4}{5} = 3\frac{1}{2}$

How to multiply fractions, p.17

$\frac{1}{2} \times \frac{1}{4} = \frac{1}{8}$ $\qquad$ $\frac{1}{2} \times \frac{1}{8} = \frac{1}{16}$

$\frac{1}{2} \times \frac{1}{6} = \frac{1}{12}$ $\qquad$ $\frac{1}{2} \times \frac{1}{3} = \frac{1}{6}$

$\frac{1}{3} \times \frac{1}{4} = \frac{1}{12}$ $\qquad$ $\frac{1}{5} \times \frac{1}{3} = \frac{1}{15}$

Multiplying fractions, p.17

$\frac{2}{3} \times \frac{3}{5} = \frac{2}{5}$ $\qquad$ $\frac{4}{5} \times \frac{3}{4} = \frac{3}{5}$

$\frac{2}{3} \times \frac{3}{4} = \frac{1}{2}$ $\qquad$ $\frac{1}{4} \times \frac{7}{8} = \frac{7}{32}$

$\frac{7}{8} \times \frac{4}{9} = \frac{7}{18}$ $\qquad$ $\frac{9}{10} \times \frac{2}{5} = \frac{9}{25}$

$\frac{3}{5} \times \frac{1}{3} = \frac{1}{5}$ $\qquad$ $\frac{2}{9} \times \frac{1}{2} = \frac{1}{9}$

$$\frac{3}{7} \times \frac{1}{5} = \frac{3}{35} \qquad \frac{4}{9} \times \frac{1}{3} = \frac{4}{27}$$

$$\frac{1}{10} \times \frac{3}{16} = \frac{3}{160} \qquad \frac{4}{13} \times \frac{2}{3} = \frac{8}{39}$$

How to divide fractions, p.18

$$\frac{1}{5} \div \frac{1}{3} = \frac{3}{5} \qquad \frac{1}{6} \div \frac{1}{5} = \frac{5}{6}$$

$$\frac{1}{3} \div \frac{2}{9} = 1\frac{1}{2} \qquad \frac{1}{9} \div \frac{1}{2} = \frac{2}{9}$$

$$\frac{2}{7} \div \frac{1}{2} = \frac{4}{7} \qquad \frac{1}{2} \div \frac{5}{7} = \frac{7}{10}$$

Dividing fractions, p.18

$$\frac{3}{5} \div \frac{1}{2} = 1\frac{1}{5} \qquad \frac{1}{9} \div \frac{3}{4} = \frac{4}{27}$$

$$\frac{3}{4} \div \frac{2}{3} = 1\frac{1}{8} \qquad \frac{8}{9} \div \frac{4}{5} = 1\frac{1}{9}$$

$$\frac{1}{3} \div \frac{5}{8} = \frac{8}{15} \qquad \frac{4}{7} \div \frac{2}{3} = \frac{6}{7}$$

$$\frac{2}{5} \div \frac{3}{8} = 1\frac{1}{15} \qquad \frac{4}{5} \div \frac{1}{2} = 1\frac{3}{5}$$

$$\frac{3}{7} \div \frac{2}{9} = 1\frac{13}{14} \qquad \frac{4}{9} \div \frac{5}{9} = \frac{4}{5}$$

Which is the bigger fraction? p.19

$$\frac{3}{4} \qquad \frac{2}{3} \qquad \frac{1}{2}$$

Simplifying fractions, p.20

$$\frac{2}{10} = \frac{1}{5} \qquad \frac{3}{9} = \frac{1}{3} \qquad \frac{5}{15} = \frac{1}{3}$$

$$\frac{6}{48} = \frac{1}{8} \qquad \frac{4}{24} = \frac{1}{6} \qquad \frac{8}{16} = \frac{1}{2}$$

$$\frac{5}{20} = \frac{1}{4} \qquad \frac{9}{45} = \frac{1}{5} \qquad \frac{7}{56} = \frac{1}{8}$$

$$\frac{16}{32} = \frac{1}{2} \qquad \frac{25}{55} = \frac{5}{11} \qquad \frac{77}{99} = \frac{7}{9}$$

Fractions problems, p.22

1. They'll have $\frac{1}{4}$ lb of cheese each.
2. If each child has 2 pieces to start with there will be 2 pieces left over. These could each be divided into 5 so each child would have $2\frac{2}{5}$ pieces altogether.
3. I must cast on 18 stitches.
4. I should use 1 lb of flour, $\frac{1}{2}$ lb fat to make double the quantity of pastry.
5. The price is correct.
6. I was three quarters of an hour late in the end.
7. Mr Ali will earn £306.80.

Which is the bigger decimal? p.25

0.2	5.3
31.4	8.6
0.6	3.79
7.8	2.8
3.64	7.1
0.05	0.006
3.11	17.623
0.2	20.1
0.54	

Write these numbers, p.25

1. 28.1
2. 32.4
3. 69.25
4. 321.3
5. 79
6. 2.05
7. 103.02
8. 97.06
9. 248.4
10. 11.079
11. 35.007
12. 345
13. 55.05
14. 18.246
15. 16.07

Multiplying & dividing decimals by 10 and by 100, p.27

$2.5 \times 10 = 25 \qquad 5.9 \times 100 = 590 \qquad 6.3 \times 10 = 63$
$1.8 \times 10 = 18 \qquad 3.5 \div 10 = 0.35 \qquad 7.8 \div 100 = 0.078$
$27.4 \times 100 = 2740 \qquad 50.3 \times 10 = 503 \qquad 45.7 \div 10 = 4.57$
$190.2 \times 10 = 1902 \qquad 70.1 \div 10 = 7.01 \qquad 640.5 \times 10 = 6405$

x 10, x 100 and x 1000, p.28

1. $10 \times 10 = 100$
2. $69 \times 10 = 690$
3. $148 \times 100 = 14,800$
4. $32 \times 10 = 320$
5. $51 \times 1000 = 51,000$
6. $78.5 \times 100 = 7850$
7. $20.25 \times 10 = 202.5$
8. $72.1 \times 100 = 7210$
9. $4.38 \times 100 = 438$
10. $16.5 \times 100 = 1650$
11. $5.94 \times 10 = 59.4$
12. $17.1 \times 10 = 171$
13. $0.06 \times 100 = 6$
14. $3.002 \times 100 = 300.2$
15. $2.15 \times 1000 = 2150$
16. $48.7 \times 100 = 4870$
17. $12.6 \times 10 = 126$
18. $192.5 \times 100 = 19,250$
19. $36.4 \times 1000 = 36,400$
20. $4.29 \times 10 = 42.9$

÷ 10, 100, 1000, p.28

1. $160 \div 10 = 16$
2. $200 \div 100 = 2$
3. $40 \div 1000 = 0.04$
4. $390 \div 10 = 39$
5. $45 \div 10 = 4.5$
6. $18 \div 10 = 1.8$
7. $98 \div 100 = 0.98$
8. $28.6 \div 10 = 2.86$
9. $35.5 \div 100 = 0.355$
10. $11.2 \div 1000 = 0.0112$
11. $120 \div 10 = 12$
12. $17.5 \div 100 = 0.175$
13. $149.6 \div 10 = 14.96$
14. $528.2 \div 10 = 52.82$
15. $39.25 \div 1000 = 0.03925$
16. $14.86 \div 10 = 1.486$
17. $439.2 \div 1000 = 0.4392$
18. $205.6 \div 100 = 2.056$
19. $78.06 \div 10 = 7.806$
20. $13.05 \div 1000 = 0.01305$

Answers

Write these in words (fractions & decimals), p.29

$\frac{3}{4} + \frac{3}{8} = 1\frac{1}{8}$ three quarters plus three eighths equals one and one eighth

$5.4 \div 9 = 0.6$ five point four divided by nine equals nought point six

$3\frac{5}{6} = \frac{23}{6}$ three and five sixths equals twenty-three over six

$3.6 = 3\frac{6}{10}$ three point six equals three and six tenths

$2\frac{6}{7} - \frac{5}{7} = 2\frac{1}{7}$ two and six sevenths minus five sevenths equals two and one seventh

$\frac{1}{10} + \frac{2}{100} = 0.12$ one tenth plus two hundredths equals nought point one two

Write these in numbers (fractions & decimals), p.29

1. $5.3 = 5\frac{3}{10}$
2. $3.34 + 0.21 = 3.55$
3. $4\frac{4}{5} - \frac{9}{10} = 3\frac{9}{10}$
4. $25 \times \frac{1}{4} = 6\frac{1}{4}$
5. $\frac{2}{3} \times \frac{3}{8} = \frac{6}{24}$
6. $22.36 - 9.7 = 12.66$
7. $\frac{1}{3} \div \frac{5}{8} = \frac{1}{3} \times \frac{8}{5}$
8. $\frac{32}{100} = 0.32$

How to add decimals, p.30

```
0.25 +        0.75 +        0.25 +
0.25          0.50          0.75
0.50          1.25          1.00

0.80 +        0.35 +        0.78 +
0.91          0.25          0.36
1.71          0.60          1.14
```

How to take away decimals, p.31

```
0.50 -        0.75 -        0.25 -
0.25          0.50          0.10
0.25          0.25          0.15

0.78 -        0.92 -        0.63 -
0.35          0.50          0.47
0.43          0.42          0.16
```

Decimals add and take away, p.32

```
5.01 +        0.80 +        2.312 +
3.56          6.07          0.041
8.57          0.90          1.206
              7.77          3.559
```

$0.48 + 6.9 + 0.07 = 7.45$
$2.06 + 3.5 + 4.44 = 10$
$12.75 + 0.9 + 8.67 = 22.32$
$0.6007 + 0.089 + 2.5 = 3.1897$

```
5.06 -     3.30 -     2.280 -     7.30 -
4.07       1.05       0.028       6.91
0.99       2.25       2.252       0.39
```

Take away the smaller number from the larger, p.32

$7 - 4.6 = 2.4$
$0.9 - 0.75 = 0.15$
$0.6 - 0.37 = 0.23$
$2.01 - 0.83 = 1.18$

$3.02 - 0.801 = 2.219$
$0.41 - 0.093 = 0.317$
$1.03 - 0.993 = 0.037$

How to multiply decimals, p33

$15 \times 0.5 = 7.5$
$152 \times 2.7 = 410.4$

$24 \times 9.8 = 235.2$
$260 \times 1.23 = 319.8$

$15.1 \times 2.3 = 34.73$
$124.2 \times 3.06 = 380.052$
$181.64 \times 0.03 = 5.4492$

$84.6 \times 0.7 = 59.22$
$25.63 \times 1.3 = 33.319$
$16.02 \times 3.11 = 49.8222$

How to divide decimals, p34

$$4.2\overline{)8.4} = 2 \qquad 2.5\overline{)12.5} = 5 \qquad 1.2\overline{)0.8412} = 0.701$$

Decimals – multiply and divide, p35

```
3.6 ×      18.3 ×     31.5 ×     16.37 ×
1.4        0.6        0.02       2.4
5.04       10.98      0.63       39.288

19.32 ×    8.007 ×    92.6 ×     7.063 ×
4.4        3.4        3.25       2.3
85.008     27.2238    300.95     16.2449
```

$$4\overline{)0.216} = 0.054 \qquad 7\overline{)0.504} = 0.072 \qquad 9\overline{)3.87} = 0.43$$

$$0.4\overline{)4.8} = 12 \qquad 0.5\overline{)3.05} = 6.1 \qquad 1.1\overline{)25.3} = 23$$

$$0.6\overline{)0.084} = 0.34 \qquad 2.1\overline{)44.1} = 21 \qquad 1.8\overline{)57.6} = 32$$

Rounding off, p36

6.88 rounds off to 6.9
5.16 rounds off to 5.2
8.92 rounds off to 8.9
9.134 rounds off to 9.1
10.026 rounds off to 10
112.55 rounds off to 112.6

Decimal problems, p.37

1. 6 litres of petrol at 84p per litre cost £5.04.
2. My bill was correct.
3. The telephone bill should be £129.41.
4. Band A units £36.2498 and Band B £13.22451.
5. £120.00 = US$219.60.
6. There are 20.12 kilometres in 12.5 miles.

Answers

Per cent, p.37

100% of £100 is £100.

50% of £10 is £5.

25% of £400 is £100.

75% of £100 is £75.

100% of 900 people is 900 people.

50% of 400 people is 200 people.

25% of 800 people is 200 people.

Percentages to fractions, p.38

$13\% = \frac{13}{100}$ $30\% = \frac{30}{100} = \frac{3}{10}$ $5\% = \frac{5}{100} = \frac{1}{20}$

$10\% = \frac{10}{100} =$ $2\frac{1}{2}\% = \frac{5}{200} = \frac{1}{40}$ $33\% = \frac{33}{100}$

$75\% = \frac{75}{100} = \frac{3}{4}$ $68\% = \frac{68}{100} = \frac{17}{25}$ $15\frac{1}{2}\% = \frac{31}{200}$

$6\frac{1}{4}\% = \frac{25}{400} = \frac{1}{16}$ $9\% = \frac{9}{100}$ $66\% = \frac{66}{100} = \frac{33}{50}$

Fractions to percentages, p.38

$\frac{3}{4} = 75\%$ $\frac{1}{10} = 10\%$ $\frac{1}{5} = 2\%$ $\frac{23}{100} = 23\%$

$\frac{3}{5} = 6\%$ $\frac{1}{20} = 5\%$ $\frac{5}{100} = 5\%$ $\frac{1}{8} = 12\frac{1}{2}\%$

$\frac{5}{8} = 62\frac{1}{2}\%$ $\frac{91}{100} = 91\%$ $\frac{17}{40} = 42\frac{1}{2}\%$

$\frac{2}{3} = 66\frac{2}{3}\%$ $\frac{1}{6} = 16\frac{2}{3}\%$

$\frac{5}{12} = 41\frac{2}{3}\%$ $\frac{3}{200} = 1\frac{1}{2}\%$

Finding percentages – practice method, p.39

50% of £40 = £20

25% of 800 = 200

75% of 1000 = 750

10% of £500 = £50

75% of 200 people = 150 people

50% of 1000 votes = 500 votes

25% of 80 countries = 20 countries

15% of 1000 cars = 150 cars

40% of 160 marks = 64 marks

Finding percentages – fraction method, p.39

5% of 900 = 45

12% of £38.00 = £4.56

32% of 525 = 168

$2\frac{1}{2}$% of £16.20 = 41p

$7\frac{1}{2}$% discount on £2,320 = £174

65 out of 650 = 10%

18 out of 30 = 60%

50 out of 250 = 20%

15p in the pound = 15%

£1.50 discount on £75.00 = 2%

Percentage problems, p.43

1. My shoes originally cost £90.
2. Alun gets £132 a week.
3. Shukri's rent has gone up to £151.20.
4. It would mean that the pound is now worth 60p by comparison with its 1981 value.
5. They're almost certain of it.
6. 'Double' means 200%.
7. A 10% deposit on £280 is £28.
8. 65.
9. No. My son got 80%.
10. My partner's pay rise is £540. My rise is £550, so I get the biggest increase.

Problems on percentage increase and decrease, p.43

1. The percentage decrease is 16%.
2. My percentage pay rise was 4.6%.
3. It cost me £32.96.
4. The new price is £124.09.
5. I would save:
   a. £5.79 on a set of pans marked £28.95.
   b. £5.84 on an electric heater marked £29.20.
   c. £5.68 on a blanket marked £28.42.
6. I will get £204.38 this week.
7. Radio sale:
   a. the cheaper radio cost £15.50.
   b. the £15.50 radio was reduced the most.

Fractions, decimals & percentages equivalents, p.47

$\frac{45}{100} = 0.45 = 45\%$ $0.67 = \frac{2}{3} = 67\%$

$\frac{20}{400} = 0.05 = 5\%$ $266\% = 2.66 = 2\frac{2}{3}$

$0.39 = \frac{39}{100} = 39\%$ $\frac{3}{5} = 0.6 = 60\%$

Calculating with ratios, p.50

1. You need 15 kilos of sand and 6 kilos of cement to make 21 kilos of mixed cement using a 5:2 mixture.
2. The cooker would be 15mm wide at that scale on the drawing.
3. I should add 400 ml of plant food to each can, i.e. 0.4 litres.
4. I won £8.75.
5. The staff:student ratio is 1:25.
6. The shop bought 168 T-shirts altogether, 56 in white and 112 in red. The shop bought 24 small, 72 medium, 48 large and 24 extra-large. Assuming the proportion of red to white T-shirts was the same for all sizes, there would be 16 small white and 8 red T-shirts, 48 white and 24 red medium T-shirts, 32 white and 16 red large T-shirts and 16 white and 8 red extra large T-shirts.
7. The ratio of day to night shift workers is 320:80 or 4:1.

Notes

Contents

Introduction

You can do the following calculations in your head, on paper, or with a calculator or other calculation aid, to check your progress.

If you find any of these difficult, take time to think about what it is that you find difficult.

If one method doesn't work for you, or you can't remember it, try another way. If you can't think of another way, look back at the relevant sections in this pack or ask a friend or a teacher. Working together often helps – find someone you can work with and explain your preferred methods to each other.

Use different methods to check your work. For example, if you did a calculation in your head, check it on paper or with a calculator.

Edit your work by checking over what you've done, especially if you make a mistake. Think of a mistake as something you can learn from, rather than something to be ashamed of.

If you're faced with awkward numbers, try doing the same calculation with easy numbers, making a note of what you do as you go along. Then try it with the original numbers again, using the same operations.

1. How much do 6 pens cost at 24p each?

2. 34×4

3. Factorise 28

4. $1\frac{1}{2}$ dozen is how many?

5. I am £35 overdrawn. How much do I need to put in my account to cover a £50 payment?

6. £1 − 47p

7. Which of these are square numbers: 24   90   25   9   84?

8. $\frac{1}{2} + \frac{1}{3}$

9. Write these as decimals, fractions and percentages: a half; a quarter; a whole one.

10. Is 5:3 the same as 15:9? Why – or why not?

11. Add 3 threes to 4 nines

12. How many minutes in 24 hours?

13. 985 divided by 11

14. Share £3.18 between 6 people

15. 328×39

16. 20% off £95

17. If the temperature was 10° Celsius during the day but dropped to −2° C overnight, how big a drop is that?

18. I spend £3.27 out of £5, how much is left?

19. How many fours in 32?

20. A third of 48

1. One wild and windy night a traveller arrived at a lonely inn. He asked the landlord for a room, explaining that he had no money with which to pay, but he did have a magic gold chain with seven links.

   The landlord was not at all sure about taking a guest with no money, even if he did have a magic gold chain, but he was a kind-hearted man, and outside the wind howled and the rain lashed down, so he asked the traveller to show him the chain.

   This the man willingly did, saying that the magic in the chain meant that it would only break in two places. He said he wanted to stay at the inn for one week and promised to pay the landlord one link every day.

   'How are you going to do that?' the landlord asked in astonishment, 'If it will only break in two places?'

   How did the traveller pay the landlord?

2. There's a pile of socks heaped together in a cupboard. There are 25 red socks and 25 green ones. If you were blindfolded, how many would you have to take out to be sure of getting a pair?

3. Twin boys stand at a fork in the road. One road leads to safety, the other road leads to danger.

   One twin always tells the truth, one twin always lies.

   What one question can you ask to be sure of finding the way to safety?

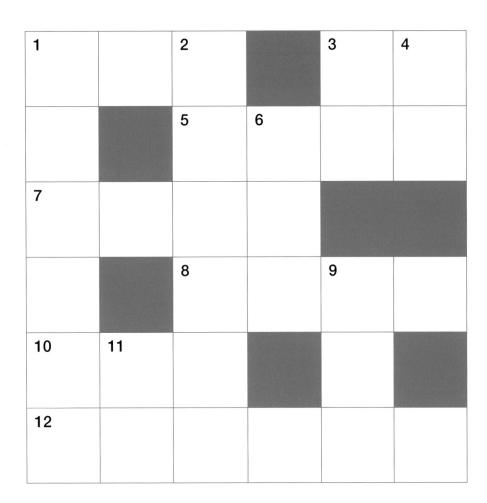

Put one number in each square.

Across

1. Three hundred and fifty minus one

3. $4\frac{2}{3}$ dozen

5. 1436 + 2078

7. 9000 + 200 + 60 + 7

8. 82 × 90

10. How many minutes in 3 hrs 33 minutes?

12. 849405 − 32875

Down

1. 524 × 647

2. How many pence in £9367.36?

3. 4 sixes plus 3 nines

4. 2 × 2 × 2 × 2 × 2 × 2

6. How many tenths in $57\frac{3}{10}$?

9. 5761 ÷ 7

11. 297 ÷ 27

Cross your numbers (2)

Put one number in each square.

Across

1. 6×6
3. $116 + 408$
5. 202×30
7. 20×10
8. 8×8
9. $21073 + 5216 + 5278$
12. 7×7
14. 12×12
16. $412 - 156$
17. thirty-one thousand, two hundred and fifteen
20. $2741 - 360$
22. six times seven
23. $3159 \div 9$
25. half of 648
27. 584×7
28. 361×24

Down

2. 892×7
3. $625 - 122$
4. 25×19
5. $712 - 85$
6. $3472 + 2773$
8. 123×5
10. $336 \div 3$
11. six thousand and fifty-two
13. $6734 \div 7$
15. half of 86
16. 2131×14
18. $941 + 97$
19. 36×20
21. a dozen
22. four elevens
24. $1\frac{1}{2}$ dozen
26. 4×10

There are **24 hours** in a day, **60 minutes** in an hour and **60 seconds** in a minute.

A quarter of an hour is **15 minutes**.

Half an hour is **30 minutes**.

Three-quarters of an hour is **45 minutes**.

90 minutes is **1 hour 30 minutes**, or an **hour and a half**.

How many hours, minutes, seconds?

Find other ways of expressing these times:

120 minutes

180 minutes

360 minutes

2 hours 25 minutes is how many minutes?

How many seconds are there in **5 minutes**?

How many hours are there in **2 days**?

What's the time?

It's three o'clock.
3.00

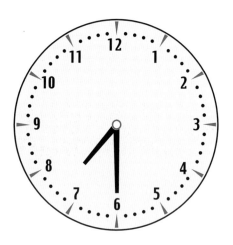

It's half past seven.
7.30

It's a quarter past twelve.
12.15

It's a quarter to six.
5.45

It's twenty past ten.
10.20

It's ten to two.
1.50

am and pm

The clock tells us what the time is but it doesn't necessarily tell us whether it's morning, afternoon, evening or the middle of the night.

Usually you can tell by common sense, if you ask somebody round to tea at four o'clock you don't expect them to wake you up at four in the morning.

For occasions when it's not so obvious what time of day is meant, we use the letters **am** and **pm**. 4 pm can only mean 4 o'clock in the afternoon.

am means morning. The letters stand for 'ante meridiem', which is Latin for before midday.

pm means afternoon and evening. The letters stand for 'post meridiem' or after midday.

10 am means 10 o'clock in the morning, **5 pm** means 5 o'clock in the afternoon.

Put these times under the right headings:

6 pm	2.55 pm	3 am
6.15 am	8 pm	9.40 pm
12 midday	3.12 am	4.30 pm

Morning	Afternoon	Evening

Write these times in figures

MSS1/L1.2

1. six o'clock

2. ten minutes past four

3. twenty-five minutes past seven

4. half past eight

5. a quarter past eight

6. half past five in the afternoon

7. quarter past eleven in the morning

8. twenty-five past ten in the evening

9. five past two in the morning

10. nine o'clock

11. twenty to five

12. quarter to eight in the evening

13. twenty-five minutes to six

14. fifteen minutes to four in the afternoon

15. five minutes to three

16. thirteen minutes to four

17. twenty-six minutes past nine

18. forty-five minutes past five

19. seventeen minutes past one

20. fifty-five minutes past eleven

Time sense

1. How long do you allow to get home from the shops?

2. How long does it take you to get to classes?

3. If I go to the launderette with a load of washing, put it in the machine and go off and do some shopping, how long should I allow before I go and pick it up?

4. You have an hour for lunch – is that long enough to go to the library, pick up your shoes from the mender's, take the dog out for a run and have your lunch?

5. Kathy leaves her little boy at the play group at 9 o'clock every morning and picks him up at 3 pm. She wants to get a part-time job locally, taking travelling time into account, what do you think are the hours she could realistically expect to do?

6. How long do you allow in the mornings to get ready to go out?

7. If you can do a journey by car that takes a quarter of an hour in the middle of the day, how long would you allow for the same journey in the rush hour?

8. How long ago was the Blitz?

9. What was the date 100 years ago today?

10. When will a baby born today be 50 years old?

11. Why do we call a century the 20th or 21st, although we say 1997, 2002 and so on?

The 24-hour clock

The 24-hour clock enables us to tell the time without having to say am or pm, in the morning, in the afternoon or so on. It's used in timetables for this reason.

The times up to midday are written in the usual way except that we put an 0 in front of the 0 to 9 so that, for example, 8 o'clock becomes 08.00 or 0800 written in the '24-hour' way.

After 12 o'clock midday we count on 12.00, 13.00, 14.00, 15.00 and so on. Five past one in the afternoon would be written 13.05 or 1305.

How would these times be written?

8.20 pm

4.15 pm

9.40 pm

3.30 am

6.30 pm

7.00 am

10.10 am

5.45 am

2.25 pm

Note: We say fourteen hundred hours for 14.00, and sixteen O five, sixteen ten for 16.05, 16.10 and so on.

The digital clock

The digital clock is used on clocks, watches, computers, videos, mobile phones, cookers and any number of other appliances.

Clock faces like those on the left are called analogue and those on the right are called digital.

Match up the times on these clocks:

Write these times as they would be shown on a digital clock:

3 pm	20 past 6
half past seven	10 to eight
quarter to nine	quarter past eleven
25 to ten	17 minutes past 4
13 minutes to 5	21 minutes to 1

Which way of showing the time do you prefer and why?

Bristol Temple Meads – Bath – London
Swansea – Cardiff – Bristol Parkway – London

MONDAY TO FRIDAYS

		WE	▲ 🍴 X	● G	🍴		🍴	■		
Swansea	d	...	...	...	0700	...	0730	...	...	0735
Neath	d	...	...	...	0710	...	0740	...	...	0750
Port Talbot Parkway	d		...	...	0719	...	0749	...	...	0801
Bridgend	d			...	0730	...	0800	...	...	0816
Cardiff Central	d	...	...	...	0755	...	0825	...	...	0855
Hereford	d	...	0703	...	0700	...	...	...	...	...
Newport	d	...	...	...	0809	...	0839	...	...	0909
Weston-super-Mare	d	...	...	0739	...	0745	...	...	0833	...
Bristol Temple Meads	d	...	...	0810	...	0840	0842	...	0910	...
Bristol Parkway	d	...	...	...	0831	...	0901	...	...	0931
Bath Spa	d	...	...	0822	...	0852	...	...	0922	...
Chippenham	d	0816	...	0835	...	0905	...	...	0935	...
Swindon	d	0834a	...	0850	0859	0920	0929	0935	0950	0959
Didcot Parkway	a	...	...	0908	...	0938	...	0951	...	1016
Oxford	a	...	0855	0941	...	1011	...	1011	...	1037
Reading	a	...	0938	0924	0927	0953	0957	1006	1019	1030
Gatwick Airport	a	...	...	1049	1049	...	1149	...	1149	...
Heathrow Term. 1	a	...	...	1025	1025	1105	1105	1105	1125	1125
Slough	a	...	1013g	1014	1014	1021	1021	1044	1053	1054
London Paddington	a	...	0947	0957	1000	1025	1027	1038	1050	1102

MONDAY TO FRIDAYS

			🍴 C	■				🍴		
Swansea	d	...	0830	...	...	...	...	0930	...	0935
Neath	d	...	0840	...	...	...	...	0940	...	0950
Port Talbot Parkway	d	...	0849	...	...	...	...	0949	...	1001
Bridgend	d	...	0900	...	...	...	...	1000	...	1016
Cardiff Central	d	...	0925	...	...	0955	...	1025	...	1055
Hereford	d	...	0819	...	...	0908	...	0937	...	...
Newport	d	...	0939	...	...	1009	...	1039	...	1109
Weston-super-Mare	d	0838	...	...	0908	...	0934	...	1028	...
Bristol Temple Meads	d	0940	0944	...	1010	...	1040	1044	1110	...
Bristol Parkway	d	...	1001	...	...	1031	...	1101	...	1131
Bath Spa	d	0952	...	...	1022	...	1052	...	1122	...
Chippenham	d	1005	...	...	1035	...	1105	...	1135	...
Swindon	d	1020	1029	1035	1050	1059	1120	1129	1150	1159
Didcot Parkway	a	1038	...	1051	...	1116	1138	...	...	1216
Oxford	a	1107	...	...	...	1142	1210	...	...	1137
Reading	a	1053	1057	1106	1118	1132	1153	1157	1219	1232
Gatwick Airport	a	...	1249	1249	1249	1331	1331	1349	1349	...
Heathrow Term. 1	a	1205	1205	1205	1225	1245	1305	1305	1325	1345
Slough	a	...	...	0947	1153	1214	1224	1224	1253	1314
London Paddington	a	1128	1130	1142	1155	1202	1221	1229	1255	1302

MONDAY TO FRIDAYS

		● 🍴		■				D		
Swansea	d	...	1030	...	...	...	...	1130	...	1135
Neath	d	...	1040	...	...	...	...	1140	...	1150
Port Talbot Parkway	d	...	1049	...	...	...	...	1149	...	1201
Bridgend	d	...	1100	...	...	...	...	1200	...	1216
Cardiff Central	d	...	1125	...	...	1155	...	1225	...	1255
Hereford	d	...	1040	...	...	...	...	1136	...	...
Newport	d	...	1139	...	...	1209	...	1239	...	1309
Weston-super-Mare	d	...	1104c	...	1104	...	1215	...	...	...
Bristol Temple Meads	d	1140	1144	...	1210	...	1240	1244	1310	...
Bristol Parkway	d	...	1201	...	...	1231	...	1301	...	1331
Bath Spa	d	1152	...	...	1222	...	1252	...	1322	...
Chippenham	d	1205	...	...	1235	...	1305	...	1335	...
Swindon	d	1220	1229	1235	1250	1259	1320	1329	1350	1359
Didcot Parkway	a	1238	...	1251	...	1316	1339	...	...	1416
Oxford	a	1307	...	...	...	1341	1403	...	...	1437
Reading	a	1253	1257	1307	1319	1332	1354	1357	1425	1434
Gatwick Airport	a	...	1449	...	1449	1549	1549	1549	1549	1649
Heathrow Term. 1	a	1405	1405	1405	1425	1445	1505	1505	1525	1545
Slough	a	...	...	1320	1354	1414	1424	1424	1453	1514
London Paddington	a	1323	1328	1340	1355	1402	1423	1429	1455	1502

X The Cathedrals Express.
Y The Red Dragon. A restaurant is available to customers joining between Swansea and Bristol Parkway.
c Change at Bristol Temple Meads and Bristol Parkway
   d = depart   a = arrive

f Change at Cardiff Central and Bristol Temple Meads.
g Change at Oxford and Reading.
r Change at Reading and Redhill.
🍴 Travelling Chef.

Reproduced with kind permission of First Great Western Co.

Use the train timetable to answer these questions:

1. When does the 08.30 from Swansea arrive at Cardiff Central station?

2. When does the 09.49 train from Port Talbot Parkway arrive in Bridgend?

3. I want to get to Oxford at about nine in the morning, to catch a connecting train. What train should I catch from Hereford to be there on time?

4. What time is the latest train you could take from Newport to get to London Paddington at around 11 am?

5. Your friend writes and asks you to meet her at Didcot Parkway off a train from Weston-super-Mare. Her train leaves Weston at 12.15, what time does it get in to Didcot?

On this timetable you read down the columns to find out when a train arrives and leaves the stations named. For example, the first train in the morning from Swansea to London Paddington leaves Swansea at 07.00 and arrives at London Paddington at 10.00.

Timetables use the 24-hour clock so 06.25 (usually written 0625 on timetables) can only be 6.25 am since 6.25 pm would be written as 18.25.

The timetable also gives extra information such as whether refreshments are available on the train. Use it to answer these questions:

6. Does the 11.40 from Neath to Swindon stop at Bath Spa?

7. Will I have to change trains if I catch the 11.55 from Cardiff Central on my way to Gatwick Airport?

8. Can I get lunch on the 11.30 Swansea to Swindon train?

9. My aunt said she'd meet me at London Paddington at 1.40 pm off the 12.35 train from Chippenham. I think she's misread the timetable, am I right?

10. Is there a restaurant on the 10.30 train from Swansea?

11. What do we mean by the 12.22 Bath Spa to London train?

12. What time does the 13.10 from Bristol Temple Meads get in to Reading?

Working with time

When I arrive at the station, the clock says **12.30**. The train I want leaves at **13.09**. How long have I got to wait?

To find out, take away the time now from the time the train leaves. There are **60** minutes in 1 hour, so there are **60 minus 30** minutes left before 13.00 hrs, that is **30** minutes. My train is at 13.09, which means I have 9 more minutes to wait after 13.00, so altogether I have to wait **30 + 9 = 39 minutes**.

It is much easier to do time calculations in your head, rather than writing them down, especially if you can imagine a clock face with the hands going round.

Adding time

My train leaves at **15.03**. The ticket collector says the journey takes **85** minutes. When should the train get in?

First write **85 minutes = 1 hour 25 minutes** (because 60 minutes = 1 hour).

Now add this time to the time the train set off. Split the calculation into two parts so that you deal with the hours and minutes separately.

Add one hour to **15.03** to give you **16.03**, then add **25 minutes** to **3 minutes** to give you **28** minutes past the hour. So the train gets in at **16.28**.

More time sense questions

1. At the pictures, the main film starts at 7.35 pm. It lasts 112 minutes. What time will it finish?

2. A train journey lasts for 3 hrs 42 min. We set off at 12.36 by the station clock. What time should we arrive?

3. The coach from London Victoria to Bristol leaves at 14.55. According to the timetable, it should arrive at 17.36. How long is the journey?

4. I went to the doctor's today. She told me to come back in 6 weeks' time for a check-up. What will the date of my next appointment be?

5. A radio announcer said there are only 152 days to Christmas. How many weeks is this?

How long does it take?

1. The news starts at 10.05 and ends at 10.15. How long is it on for?

2. London to Birmingham is 118 miles. My average speed for the journey is 50 mph. If I leave home at 1.30pm, what time do I arrive in Birmingham?

3. My train leaves Waterloo at 8.24 am. The journey takes 20 minutes. At what time do I arrive in Richmond?

4. Tom works in a shop from 6.00 pm to 7.45 pm each weekday evening and from 9.00 am to 1.00 pm on Saturday. He gets £5.50 per hour.

   a) How many hours does he work during the week in the evenings?

   b) How many hours does he work on Saturdays?

   c) How many hours does he work altogether?

   d) How much money does he get altogether for a week's work?

5. Jolene is a garage attendant every evening from 8.00 pm to 1.15 am. She works from 5.00 pm to 2.00 am on Saturday.

   a) How many hours does she work in the evenings during the week?

   b) How many hours does she work on a Saturday night?

   c) She is paid £4.50 per hour. How much does she earn in a week?

6. My train leaves at 22.56 and arrives at its destination at 23.40. How long does the journey take?

7. A film lasts from 8.35 to 10.05. How long is it on for?

8. A radio programme lasts for $1\frac{1}{2}$ hours. It starts at 10.20 am. When does it finish?

9. My uncle was 63 in 1998. When was he born?

10. My sister was born in 1986. How old is she now?

11. Ben was born in 1952. When can he retire?

12. The voyage from England to Norway by sea takes 22 hours. If the boat leaves at 15.30 on Tuesday, when does it arrive in Norway?

The calendar

The calendar helps you find your way around the year, just as a map helps you find your way around an area.

To use a calendar you need to know the months of the year and the days of the week. It helps to know the seasons too. In the northern hemisphere midwinter is in December and midsummer is in June. In the southern hemisphere midwinter is in June and midsummer is in December. The chart below shows the seasons in the northern hemisphere.

The months	The seasons
January February	Winter
March April May	Spring
June July August	Summer
September October November	Autumn
December	Winter

The days of the week are Monday, Tuesday, Wednesday, Thursday, Friday, Saturday, Sunday.

Now answer these questions:

1. How many months are there in a year?

2. How many days are there in a week?

3. What are the names of the seasons?

4. What is the month after May?

5. What is the month after December?

6. What is the date today?

7. What season are we in?

8. What is the hottest time of the year in Europe? Is that the hottest time in Australia?

Writing the date

The quick way to write the date is to put the number of the day, the number of the month and then the last two numbers of the year (so long as it's clear which century is meant, otherwise write the year in full).

The number of the month could be one or two figures. For example, February can be represented by 2 or 02.

Written this way the **2nd June 2002** becomes **2.6.02 (or 2/6/02)** the **2nd** day of the **6th** month of the **2nd** year of the twenty-first century.

You can write either 2nd June 2002 or June 2nd 2002 but it is important to put the number of the day first when you are writing the date the quick way.

For example, **2nd June 2002** would be **2.6.02**, but **6.2.02** would be the 6th February 2002. Just to make life difficult, in the United States of America and in some other countries they do put the number of the month first!

Write these dates the quick way:

January 3rd 1999 ....

4th June 2003 ....

September 10th 1962 ....

May 21st 1942 ....

17 October 1951 ....

Write these dates in full:

3.8.66 ....

16.10.03 ....

30/09/77 ....

29.12.47 ....

22.2.70 ....

1/5/81 ....

Days of the month

Thirty days has September,
April, June and November.
All the rest have thirty-one,
Except for February alone,
Which has twenty-eight days clear,
And twenty-nine in each Leap Year.

Now answer these questions by filling in the gaps:

1. How many days are there in October?

   There are .... days in October.

2. How many days are there in June?

   There are .... days in June.

3. How many days are there in September?

   There are .... days in September.

4. How many days are there in February?

   There are usually .... days in February.

5. How many days are there in July?

   There are .... days in July.

The calendar – abbreviations

The months:

Jan is short for _January_

Feb is short for ....

Mar is short for ....

Apr is short for ....

May is written in full as ....

June is written in full as ....

July is written in full as ....

Aug is short for ....

Sept is short for ....

Oct is short for ....

Nov is short for ....

Dec is short for ....

The days:

Mon is short for ....

Tues is short for ....

Weds is short for ....

Thurs is short for ....

Fri is short for ....

Sat is short for ....

Sun is short for ....

Dates and ages: personal

1. When were you born?

   I was born in ....

2. What is your date of birth?

   My date of birth is ....

3. How old are you now?

   I am .... years old.

4. How old were you in 1990?

   In 1990 I was ....
   years old.

5. How old will you be in 2010?

   I will be .... years
   old in 2010.

Multiply your age by 7. Multiply the
answer by 1443.

What do you notice about the answer?

Using the calendar

1. My friend's birthday is in three weeks
   time, on a Thursday. What date is
   that?

2. I've booked my holiday for two
   weeks in July. How many months
   away is it?

3. Gina's little girl is exactly one month
   older than my son. His birthday is on
   the 5th of this month so when is it
   Gina's daughter's birthday?

4. When is midsummer's day?

5. Mrs Green has a dental appointment
   at 10.30 on Monday morning.
   What date will that be?

6. I made some wine which is supposed
   to be kept for six months before you
   drink it. I made it in June, is it ready
   yet?

7. My driving test is on 12th October.
   How long have I got to get ready
   for it?

8. Amina comes to evening class twice
   a week on Tuesdays and Thursdays.
   In a month's time she goes on
   holiday, what is the date of her last
   Thursday class before she goes?

9. When do we put the clocks forward
   and backwards? Why do we do it?

10. There are two days in the year when
    the day and night are the same
    length. When are they?

11. Bill was born on the 29th February
    1960. How old is he now? How many
    times has he been able to celebrate
    his birthday on the right day?

12. Why do you think we have a leap
    year every four years?

Ages

Fred's sister Julie is 24, she is 5 years older than him. How old is Fred?

How old will Julie be in seven years time?

How old was Fred three years ago?

What will Julie's age be when Fred is 24?

What year was Fred born?

Fred has four brothers and one sister. Their ages are 21, 14, 17, 25 and 24. What is the average age of everyone in Fred's family (including Fred)?

What year was the youngest born?

What is the difference in years between the youngest and the eldest in Fred's family?

N1/L1.3, HD1/L1.3

Dates and ages

What year is this?

When were you born?

How old are you?

To find the age of a person when you know the year they were born:

2004 −
1982   this year
         year born
         take away to find the age
         of the person.

To find the year of birth when you know the age of the person:

2004 −
36   this year
         age of person
         take away to find the year of
         birth.

Find the ages of people born in these years:

1. 2003 ....   5. 1947 ....

2. 1970 ....   6. 2000 ....

3. 1956 ....   7. 1968 ....

4. 1932 ....

When were people of these ages born?

1. 9 years....................   5. 2 years....................

2. 21 years....................   6. 17 years....................

3. 54 years....................   7. 25 years....................

4. 30 years....................

Pounds and pence

We use 2 units of money, pounds **(£)** and pence **(p)**

There are **100 pence in one pound**, so, 200p = £2, 500p = £5, 1000p = £10 and so on.

Whole pounds are written like this: £1, £2, £10, £18, £52, £150, and so on.

Pounds and pence are written in decimal form with whole pounds, decimal point, then 2 figures to show pence from 1p to 99p.

> **£1.25** is one pound twenty-five pence.
>
> **£2.86** is two pounds eighty-six pence.
>
> **£1.05** is one pound five pence.
>
> **£1.50** is one pound fifty pence.
>
> **£3.09** is three pounds nine pence.

Note that for 1p to 9p over the pound we write 01, 02... 08, 09 after the point to avoid confusion with 10p, 20p... 80p, 90p.

Whole pounds can also be written with zero pence: £1.00, £2.00, £10.00.

How many pence?

£1 = 100p so **£2 = 200p £3 = 300p** and so on.

How many pence in **£1.67**?

    100p in £1 + 67p extra.   100p + 67p = **167p**

How many pence in **£3.42**?

    In £3 there are 3 × 100p = 300p, plus 42p extra.   300p + 42p = **342p**

How many pence in **£11.06**?

    In £11 there are 11 x 100p = 1100p, plus an extra 6p. 1100p + 6p = **1106p**

So:
                    £1.67 = 167p
                    £3.42 = 342p
                    £11.06 = 1106p

The rule is, ignore the point, write the same numbers and call them pence.

How many pounds?

How many pounds in **250p**?

There are 100 pence in £1, so 200p is £2. That leaves 50 pence.

So 250p = **£2.50**

How many pounds in **536p**?

Count the hundreds: 500p is £5. That leaves 36p.

So 536p = **£5.36**

The rule is: count 2 numbers from the right, put in the decimal point and call it pounds.

So 1234p = **£12.34**   199p = **£1.99** and so on.

Less than 100 pence

45p is less than £1. To write 45p as pounds we have to put 0 in front of the point because there are no whole pounds:

45p = £0.45

Any amount between 1p and 99p can be written this way.

Look at 'Place value: decimals' in Book 3 for more on this.

Change these sums of money to pounds:

55p ....

88p ....

101p ....

172p ....

2566p ....

430p ....

Giving change and getting change

Do you know how the shop assistant works out what change to give you?

Can you check your change to see if it's right?

If you've ever looked at the coins in your hand and thought, 'I'm sure I should have more than that' read on…

Suppose you're in a shop and you spend 42p and pay with a 50p piece. The assistant gives you 8p change. He or she may count the coins into your hand like this:

"42, 43, 45, 50"

Can you work out what coins you would have been given?

The point is that the assistant used the coins to count on, one for a penny, two for a two-pence piece, five for a five-pence piece and so on.

Try these questions, using the counting on method and real money if possible:

1. What change would you get from £1 if you spent 81p?

2. What change would you get from £5 if you spent £3.48?

3. I spent 83p in the newsagent's, paid with a £1 coin and got 13p change. Was that right?

4. My bill came to £14.03 in the supermarket. I gave the assistant three five pound notes and three pence. What change should I have got?

5. At the butcher's my bill came to £12.67. They were short of silver in the shop, so the assistant asked me: 'Have you got the 17p?'. I gave it to him with a twenty pound note, and he gave me my change. What was it?

6. Take a handful of coins from your pocket or your purse and see if you can guess how much money there is. Check your answer and see how close your guess was.

MSS1/E2.2

```
        D.S.
      NEWSAGENTS

    17 MAY 2004

      01.99
      00.31
      01.23
      00.95


      04.48  CASH
```

```
        C.B.
      SUPERMARKET

001.29
000.27
001.46
000.29
001.75
002.08
007.14  TOTAL
008.00  CASH  TENDERED
000.86  CHANGE
```

1. What does the bill come to?

2. I gave the cashier a £5 note. How much change did I get?

3. Write the date at the top of the list of prices in figures.

1. What does the bill come to?

2. How much did I give the cashier?

3. How much change did I get?

4. Did I get the right amount of change?

Shopping – find the cost

1. 1 kg of apples cost 90p, so:

   a) 2 kg of apples cost ....

   b) 3 kg of apples cost ....

   c) $\frac{1}{2}$ kg apples cost ....

   d) $1\frac{1}{2}$ kg apples cost ....

2. 1 packet of butter costs £1.09, so:

   a) 2 packets of butter cost ....

   b) 4 packets of butter cost ....

3. If 1 packet of butter weighs 250g, how much do:

   a) 2 packets weigh ....

   b) 4 packets weigh ....

4. 125g of paté costs 76p, so:

   a) 250g of paté costs ....

   b) half a kilo of paté costs ....

5. 100g of bacon costs 65p, so:

   a) 200g of bacon costs ....

   b) 400g of bacon costs ....

   c) 300g of bacon costs ....

How much does it cost?

1. 3 lbs of apples at 55p per lb.

2. 7 metres of cotton at £1.80 per metre.

3. 5 metres of rope at £4.45p per metre.

4. 6 yards of woollen cloth at £15 per yard.

5. 4 lbs of onions at 35p per lb.

6. 6 packets of envelopes at 99p per packet.

7. 8 pints of milk at 66p per litre.

8. 7 tins of dog food at 45p per tin.

9. 4 tubes of toothpaste at £1.77 per tube.

10. 8 cups of tea at 80p per cup.

11. 5 pints of beer at £2.30 per pint.

12. 3 paperback books at £7.99 per book.

13. 7 newspapers at 45p per newspaper.

14. 3 dozen eggs at £1.85 a dozen.

15. 2 pairs of tights at £3 per pair.

How much change do you get?

1. At the butcher's your meat costs £9.62. You pay with a £10 note.

2. At the baker's your bill comes to £1.54. You pay with two £1 coins.

3. At the supermarket your bill comes to £27.92. You pay with two £20 notes.

4. At the chemist's you need to pay £4.63. You pay with a £5 note.

5. At the vegetable stall your fruit and vegetables come to £3.76. You pay with a £5 note.

6. At the delicatessen your bill comes to £11.23. You pay with one £10 note and a £5 note.

7. At the stationer's your goods come to £17.45. You pay with a £20 note.

8. At the fish and chip shop your bill comes to £6.54. You pay with two £5 notes.

9. At the restaurant your bill comes to £26.77. You pay with two £10 notes, a £5 note and two £1 coins.

10. At the railway station your ticket costs £3.25. You pay with four £1 coins.

How much does it cost? (2)

1. A driving school charges £12.10 per lesson. How much will 15 lessons cost?

2. 1 litre of petrol costs 87p. How much will 6 litres cost?

3. A meal at a restaurant costs £25.00 per person. How much will it cost if Mr and Mrs Patel, Mrs Patel's brother and his partner all go for a meal together?

4. A seat at the cinema costs £6.85. If 7 people go together how much will they pay?

5. One kitchen chair costs £26.99. I need four chairs. How much will I have to pay?

6. A man bought 1 kg of onions at £2.29 per kg, 1 kg of carrots at 59p per kg, and 300 g of mushrooms at 24p per 100 g. How much did he pay altogether?

7. A daily newspaper costs 40p, and it's not published on Sunday. How much did I pay the newsagent for a fortnight's papers?

8. A film costs £7.99 to develop. How much will 7 films cost?

9. My rent is £127.00 per week. How much per year?

10. 250 kg of butter costs £1.05. How much will 3 kg cost?

11. 50 g of wool costs £1.85p. How much will 350 g cost?

12. Whitechapel to Oxford Circus costs £2.60 on the tube. How much will it cost to travel to Oxford Circus and back once a day for 4 days?

13. A woman is paid £8.60 per hour. She works 40 hours per week. What is her weekly wage?

14. My television rental is £9.40 per month. How much do I pay in a year?

15. London to Swansea is 193 miles. My car does 6 miles per litre. How many litres of petrol must I buy for a trip there and back?

16. I buy 5 and a half metres of net curtain material for £1.82 a metre. How much does it cost?

MSS1/E2.2, N2/L1.5

Money sense

1. I've got £20 in my pocket and I want to stock up at the supermarket. Can I afford to get rice, flour, sugar, tea and coffee? How much would you expect to pay for each item? What quantity of each item would you be able to get?

2. What are the things you have to pay for every week, including household bills such as rent, as well as food. How much do you reckon you need to keep going?

3. What are the things you pay for every three months (quarterly)?

4. What is the current rate of inflation, what does it mean?

5. A 250g jar of instant coffee is marked at £4.65. Would I be better off buying two 125g jars at £2.40 each?

6. I want to buy a digital television and the one I like costs £450 if I pay cash, or £50 down and 12 monthly payments of £40 if I buy it through an extended credit scheme. How much would I save if I could afford to pay cash? Which way would you choose to buy it? Why?

7. What is meant by the term APR?

8. What do the letters VAT stand for? What is the current rate of VAT?

9. What is the current base rate? What does the term 'base rate' mean?

10. What is meant by the Financial Times Share Index?

Currency conversion

Converting from one currency to another is essential if you are going to visit another country to work, on holiday or to visit family or friends.

The conversion rate is the value of one currency compared with another. For example, £1 (one pound) may be worth €1.50 (1 Euro 50 cents) and €1 may be worth £0.67 (or 67p).

You can do currency conversions yourself, using a calculator, on paper, or in your head, or you can use one of the currency converters on the Internet.

On the Internet, enter the amount you want to convert and your chosen currency. The calculation is done automatically and the conversion rate is constantly updated.

Some personal organisers (also known as personal digital assistants, PDAs) have a currency conversion facility which works in much the same way.

If you are abroad and you want to know roughly how much things cost, find an equivalent amount which you can use as a mental guide. For example, if there are **2.4** Australian dollars to the pound sterling, then roughly speaking, **10** Australian dollars will be worth about **£4**.

If the Euro is worth **67p**, then each Euro is worth roughly two thirds of a pound. If the Euro was worth **73p** it would be nearly three quarters of a pound. So, if you see something on sale at €**300**, that's **£201** if one Euro is worth **67p**, or **£219** if one Euro is worth **73p**.

To get an accurate conversion based on the value of €**1** = **67p**, multiply **300** by **0.67**.

Another way of doing this calculation, which avoids having to use decimals, is to say '**3 times 67**'. You can do this because **300** × **0.67** can be written as **3** × **100** × **0.67**. Then if you multiply **0.67** by **100** you get **67**, which leaves you with **3** × **67**.

MSS1/L2.1

Driving

1. I drive from London to Leeds.

   In London, my mile counter reads 65403.

   In Leeds, it reads 65601.

   How far is it from London to Leeds?

2. Before I start, I fill up with petrol.

   I buy 26 litres at 83p. How much altogether?

   I also buy 5 litres of oil at £6.87, and 2 bars of chocolate at 43p. What is my total bill?

3. Half way there, I stop for a rest.

   How many miles have I travelled?

   What does the mile counter read?

4. I arrive at 6.15pm.

   I had set off at half past two.

   How long was the journey?

   Make a rough guess at my average speed and then work it out accurately.

Problems

1. Dave has £15 to spend.

   He buys a CD for £11.50, and a magazine for £2.50p.

   How much has he left?

2. I buy 6 pieces of wood, costing as follows: 2 at £3.60; 2 at £3.00; and 2 at £2.70. VAT at 17.5% was added. I was charged £24.86, was this correct?

3. A train journey lasts 4 hours 20 minutes.

   We leave the station at 12.15pm, what time should we arrive?

   In fact, there is a delay of 25 minutes.

   When do we actually arrive?

4. A bottle of wine contains 70 centilitres.

   How many 125ml glasses of wine will you get from the bottle?

The metric system

We use the metric system to measure length, weight (or mass) and capacity (or volume). The metric system is based on the number 10, in the same way as the decimal system. Unlike imperial (English) units, one system covers length, weight and capacity. Calculating is easier in the metric system, once you know how it works.

The metric unit of length is the **metre** (m)

The metric unit of weight is the **gram** (g)

The metric unit of capacity is the **litre** (l)

Each unit has a two-part name which consists of one of the following:

kilo, hecta, deca, deci, centi or milli followed by a metre, gram or litre.

Kilo is the biggest, milli is the smallest. This is how they fit together:

kilometre km	hectometre hm	decametre dam	metre m	decimetre dm	centimetre cm	millimetre mm
kilogram kg	hectagram hg	decagram dag	gram g	decigram dg	centigram cg	milligram mg
kilolitre kl	hectalitre hl	decalitre dal	litre l	decilitre dl	centilitre cl	millilitre ml

Biggest   × 10   ÷ 10   Smallest

× 100   ÷ 100

× 1000   ÷ 1000

Each unit is 10 times bigger than the one to the right.

Each unit is 10 times smaller that the one to the left.

kilo means 1000 times bigger	milli means 1000th part
hecta means 100 times bigger	centi means 100th part
deca means 10 times bigger	deci means 10th part

Not all the units are in everyday use. The ones we use most are outlined boldly in the table above. They are:

length	–	metre m, centimetre cm, millimetre mm, kilometre km
weight	–	gram g, milligram mg, kilogram kg (known as a 'kilo')
capacity	–	litre l, centilitre cl, millilitre ml.

Very large weights are measured in **tons** (sometimes called metric tons).

1 metric ton = 1000 kilograms

MSS1/L1.4

Going metric

In the last 30 years Britain has been changing from a system of measuring called the imperial system to the metric system.

This has meant that many of the items that you buy are no longer sold in the old measurements such as pounds and ounces, yards, feet and inches, gallons and pints. They are now sold in metric measurements such as grams and kilograms, metres and centimetres, litres and millilitres.

Note: we say 'kilo' instead of kilogram, but all the other metric measures are said in full.

The change from the old system to the metric system is still not finished. You can still see road signs with distances measured in miles rather than kilometres. When you buy drinks in a pub, they are still sold in pints rather than litres. It has become illegal to sell most food items in imperial weights, although prices are often displayed in both metric and imperial systems, i.e. in lbs and kgs.

Many people still measure their height and weight in feet and inches, and stones and pounds, rather than in metres and kilograms.

The two systems of measurement are just two different ways of saying the same thing.

Generally, it's not necessary to convert from one system to the other, it's far better to 'think metric', though this is easier said than done if you have grown up using the old measurements.

It may help you to get a rough idea of the way the two systems relate if you remember these sayings:

> A metre measures three foot three, it's longer than a yard, you see.
>
> A litre of water's a pint and three quarters.
>
> Two and a quarter pounds of jam weigh about a kilogram.

If you want to convert from one system to the other exactly you need to use a conversion table like the one in this section, or a calculator.

Metric conversion tables

These tables can be used to convert measurements from imperial to metric and from metric to imperial. The bold figure in the centre column can be used to convert the measurement on either side, e.g. 1 in = 2.54cm and 1 cm = 0.39 in; 9 yds = 8.23 m; 9m = 9.84 yds.

Length

cm		in		metres		yards
2.54	1	0.39		0.91	1	1.09
5.08	2	0.79		1.83	2	2.19
7.62	3	1.18		2.74	3	3.28
10.16	4	1.58		3.66	4	4.37
12.70	5	1.97		4.57	5	5.47
15.24	6	2.36		5.49	6	6.56
17.78	7	2.76		6.40	7	7.66
20.32	8	3.15		7.32	8	8.75
22.86	9	3.55		8.23	9	9.84
25.40	10	3.94		9.14	10	10.94
50.80	20	7.88		18.29	20	21.87
76.20	30	11.82		27.43	30	32.81
101.60	40	15.76		36.58	40	43.74
127.00	50	19.70		45.72	50	54.68
152.40	60	23.64		54.86	60	65.62
177.80	70	27.58		64.01	70	76.55
203.20	80	31.52		73.15	80	87.49
228.60	90	35.46		82.30	90	98.42
254.00	100	39.40		91.44	100	109.36

Distance

km		miles
1.61	1	0.62
3.22	2	1.24
4.83	3	1.86
6.44	4	2.49
8.05	5	3.11
9.66	6	3.73
11.27	7	4.35
12.87	8	4.97
14.48	9	5.59
16.09	10	6.21
32.19	20	12.43
48.28	30	18.64
64.37	40	24.86
80.47	50	31.07
96.56	60	37.28
112.65	70	43.50
128.74	80	49.71
144.84	90	55.93
160.93	100	62.14

Metric conversion tables – continued

Weight

kg		pounds (lb)	grams		oz
0.45	**1**	2.20	28.35	**1**	0.04
0.91	**2**	4.41	56.70	**2**	0.07
1.36	**3**	6.61	85.05	**3**	0.11
1.81	**4**	8.82	113.40	**4**	0.14
2.27	**5**	11.02	141.75	**5**	0.18
2.72	**6**	13.23	170.09	**6**	0.21
3.18	**7**	15.43	198.44	**7**	0.25
3.63	**8**	17.64	226.79	**8**	0.28
4.08	**9**	19.84	255.14	**9**	0.32
4.54	**10**	22.05	283 49	**10**	0.35
9.07	**20**	44.09	311.84	**11**	0.39
13.61	**30**	66.14	340.19	**12**	0.42
18.14	**40**	88.18	368.54	**13**	0.46
22.68	**50**	110.23	396.89	**14**	0.49
27.22	**60**	132.27	425.23	**15**	0.53
31.75	**70**	154.32	453.58	**16**	0.56
36.29	**80**	176.36	566.98	**20**	0.71
40.82	**90**	198.41	850.47	**30**	1.06
45.36	**100**	220.45	1133.96	**40**	1.41
			1417.45	**50**	1.77
			2834.90	**100**	3.53

Area

sq m		sq yds
0.84	**1**	1.20
1.67	**2**	2.39
2.51	**3**	3.59
3.34	**4**	4.78
4.18	**5**	5.98
5.02	**6**	7.18
5.85	**7**	8.37
6.69	**8**	9.57
7.52	**9**	10.76
8.36	**10**	11.96
16.72	**20**	23.92
25.08	**30**	35.88
33.44	**40**	47.84
41.81	**50**	59.80
50.17	**60**	71.76
58.53	**70**	83.72
66.89	**80**	95.68
75.25	**90**	107.64
83.61	**100**	119.60

Capacity

litres		pints	litres		gallons
0.57	1	1.76	4.55	1	0.22
1.14	2	3.52	9.09	2	0.44
1.70	3	5.28	13.64	3	0.66
2.27	4	7.04	18.18	4	0.88
2.84	5	8.80	22.73	5	1.10
3.41	6	10.56	27.28	6	1.32
3.98	7	12.32	31.82	7	1.54
4.55	8	14.08	36.37	8	1.76
5.11	9	15.84	40.91	9	1.98
5.68	10	17.60	45.46	10	2.20
11.36	20	35.20	90.92	20	4.40
17.05	30	52.79	136.38	30	6.60
22.73	40	70.39	181.84	40	8.80
28.41	50	87.99	227.30	50	11.00
34.09	60	105.59	272.76	60	13.20
39.78	70	123.19	318.22	70	15.40
45.46	80	140.78	363.68	80	17.60
51.14	90	158.38	409.14	90	19.80
56.82	100	175.98	454.60	100	22.00

Use the tables to work out these conversions:

1. How many metres in 10 yards?

2. How many pounds in 5 kilograms?

3. How many yards in 100 metres?

4. How much do you weigh in kilograms?

5. How tall are you in metres?

MSS1/L1.4, MSS1/L2.6

Units of measurement

Write down some things that we used to measure in:

- yards
- pounds (lb)
- pints
- inches
- gallons
- ounces
- miles
- tons

We now measure most things in metric units.

The most commonly-used metric units are:

- millimetres mm
- metres m
- kilometres km
- litres l
- centilitres cl
- centimetres cm
- grams g
- kilograms kg
- millilitres ml

Which measure length? ....

Which measure quantities of liquid? ....

Which measure weight? ....

Which is bigger:

a metre or a centimetre? ....   a litre or a centilitre? ....

a gram or a kilogram? ....   a millilitre or a centilitre? ....

a millimetre or a metre? ....

Fill in the missing numbers:

1. There are .... centimetres in one metre   .... cm = 1 m

2. There are .... centilitres in one litre   .... cl = 1 l

3. There are .... metres in one kilometre   .... m = 1 km

4. There are .... grams in one kilogram   .... g = 1 kg

5. There are .... millimetres in one metre   .... mm = 1 m

6. There are .... millimetres in one centimetre .... mm = 1 cm

7. There are .... millilitres in one litre   .... ml = 1 l

> **Look at these words:**
>
> century
>
> cent
>
> percentage
>
> centigrade
>
> centipede
>
> centenary
>
> Which number do they all refer to? Explain how.

Look again at your list of things.

Write next to them the metric units we measure them in now.

Which metric units would you use to measure these things? (Use a tape measure to help you with some and collect the labels from things you buy.)

screws	the distance from Cardiff to Bristol
knitting wool	petrol
curtain material	the height of a building
speed of a car	bottle of lemonade
milk	your weight
spanners	how much water a bucket can hold
sugar	clothes size
a spoonful of medicine	waist measurement
tea	potatoes
the size of a piece of paper	how much sand a bucket can hold

MSS1/L1.4

Weights and measures (imperial)

Fill in the gaps:

1. There are .... ounces in 1 lb.

2. There are .... inches in 1 foot.

3. There are .... fluid ounces in 1 pint.

4. There are .... feet in 1 yard.

5. There are .... inches in 1 yard.

6. There are .... lbs in 1 stone.

7. There are .... yards in 1 mile.

8. There are .... pints in 1 gallon.

MSS1/L1.7

Weights and measures (metric)

Fill in the gaps:

1. There are .... grams in 1 kilogram.

2. There are .... centimetres in 1 metre.

3. There are .... millilitres in 1 litre.

4. There are .... metres in 1 kilometre.

5. There are .... milligrams in 1 gram.

6. There are .... litres in 1 kilolitre.

7. There are .... millimetres in 1 metre.

8. There are .... centilitres in 1 litre.

9. There are .... decilitres in 1 litre.

10. There are .... kilograms in 1 metric ton.

MSS1/L1.4, MSS1/L1.7

Measure these lines

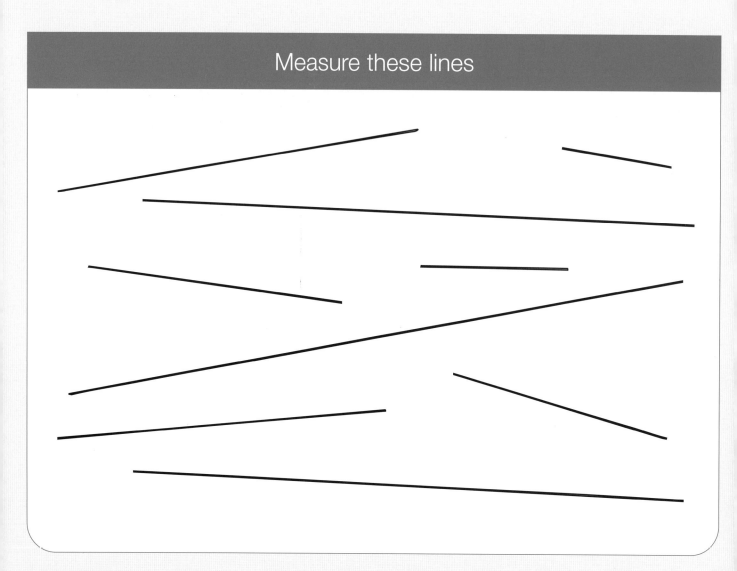

MSS1/E2.5, MSS1/E3.5

The table top

Your book

Your pen or pencil

The door of the room

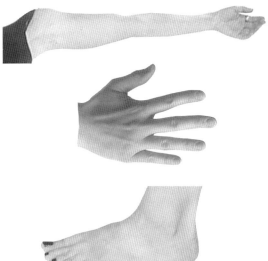

The length of your arm

The span of your hand

The length of your foot

MSS1/E3.5

Vital statistics – women

waist   shoe size   weight

height   bust   hips

Height ....   Waist ....

Weight ....   Hips ....

Bust ....   Shoe size ....

Vital statistics – men

height   weight   shoe size   chest

inside-leg   collar size   waist   hips

Height ....   Waist ....

Weight ....   Hips ....

Collar size ....   Inside-leg ....

Chest ....   Shoe size ....

MSS1/E3.5, MSS1/E3.6

Weight – working out your body mass index

Body mass index is used to determine whether or not an individual's weight is within the range considered to be healthy.

To work out your body mass index:

1. measure your height in metres and multiply the figure by itself (i.e. square your height)

2. weigh yourself in kilograms

3. divide your weight by your height squared.

For example, if you are 1.6m (5 feet 3 inches) tall and weigh 65kg (10 stone), this is the calculation you would do:

Square your height: $1.6 \times 1.6 = 2.56$

Divide your weight by your height squared: $65 \div 2.56 = 25.39$

The answer is your BMI: 25 (correct to the nearest whole number).

These are the recommended BMI categories for a healthy adult, as set out by the government's Food Standards Agency:

below 20	**underweight**
20–24.9	**ideal**
25–29.9	**overweight**
30–39.9	**obese**
over 40	**extremely obese**

BMI is only a guideline and it's only one measure of your health. Body fat percentage, blood pressure, resting heart rate, cholesterol and other measurements are also important.

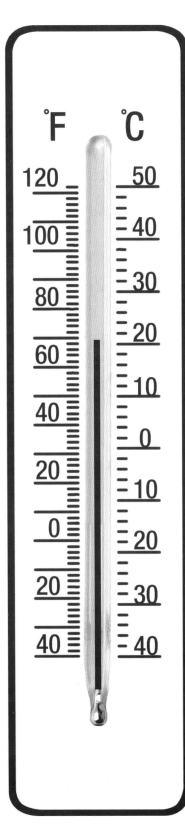

The scale on a thermometer is marked in degrees Celsius (or centigrade) and Fahrenheit.

To find out what the marks mean on the scale count the spaces in between two marked numbers.

For example, on the Fahrenheit scale there are 5 spaces to show 10 degrees. So each space between the marks is 2 degrees.

On the Celsius or centigrade scale there are 10 spaces to show 10 degrees, so each space between the marks is 1 degree.

The freezing point of water is 0°C or 32°F.

The boiling point of water is 100°C or 212°F.

Temperatures below 0°C are written with the negative sign as −1°C (minus one degree Celsius) −10°C and so on. These may be referred to as 'degrees of frost'.

Match them up

oz	stone	in	square feet
km	foot	g	miles per hour
yd	centimetre	hp	gallons
lb	yard	am	miles per gallon
ft	pound	pm	horse power
cm	metre	mph	square centimetres
st	kilogram	sq ft	morning
kg	kilometre	cm²	afternoon and evening
m	ounce	gal	inch
		mpg	grams

%	minus, take away
+	pound
=	divided by
−	plus, add
×	inch
÷	per cent
"	equals
'	multiplied by, times
£	feet

$\frac{1}{2}$	a half	pt	minutes
$\frac{3}{4}$	a quarter	tsp	regulo (gas)
$\frac{3}{8}$	a third	tbsp	litre
$\frac{1}{3}$	three quarters	°C	hours
$\frac{1}{8}$	five eighths	°F	teaspoon
$\frac{1}{4}$	an eighth	l	tablespoons
$\frac{5}{8}$	three eighths	fl oz	degrees Celsius/ centigrade
•	and	reg	fluid ounces
( )	decimal point	hr	degrees Fahrenheit
&	brackets	min	pints

mm	cubic feet
sq in	metre
cc	square inch
sq yd	cubic centimetre
cu ft	square yard
m	millimetre
km	litre
l	kilometre

N2/E3.1, MSS1/L1.4

It is important to have a rough idea of what the answer will be before you start, then you will realise if your answer is wildly wrong, and you can check it.

This is particularly important when calculating with money or any other numbers with a decimal point. Make sure you get the decimal point in the right place, or your answer could be 10 or 100 times too big or too small.

If you're doing a calculation with several stages and you want a grand total at the end, use the memory button on the calculator to store your answers as you go along and then add them all at the end using the M+ button.

Try these sums with decimals:

$£12.50 + £306.20 =$ $48.02 \times 13 =$

$£46.23 - £13.95 =$ $11.65 \times 7 =$

$£60.00 \times 8 =$ $98.03 + 16.5 =$

$£96 \div 12 =$ $12.01 \div 3.2 =$

$15 \times £28.06 =$ $£45.02 + £19.36 =$

$£2.50 - £0.75 =$ $£890.07 + £96.90 =$

Knowing what calculation to do

This is one of the most important skills to develop, it's no good knowing how to operate the calculator but not knowing what calculation you ought to be feeding into it. Use common sense and don't panic and you'll find that most of the problems you come up against in everyday life that can be solved mathematically are nowhere near as complicated as they look. Think clearly, ignore any unnecessary information and try to see what is really important and what is just confusing.

For example, suppose you decide to buy a new carpet for the living room so you go to the show room and see dozens of different carpets all at different prices and of different qualities. Decide which two or three carpets you really do prefer, then work out how much each would cost you with the aid of the next exercise.

Carpeting a room

The questions you need to think about are:

How much carpet do you need and how much will it cost?

Using a calculator will make the sums easier.

First, measure the room. Draw a rough plan and mark the measurements from wall to wall, using a metre stick or a tape measure.

Say the room looks like this:

The length is 5m 15cm, or 5.15m

The width is 3m 24cm, or 3.24m

Now find the **area of the floor** by multiplying the length by the width. This tells you how much carpet you will need. The answer will be in square metres.

First make a guess by rounding off the measurements to whole numbers.

Convert the measurements to metres, as carpet is measured and priced by the square metre.

Make a rough guess at the answer: $5 \times 3 = 15$

Find the exact answer, using a calculator if you prefer:

$$5.15 \times 3.24 = 16.686$$

This figure is **square metres** (metres x metres, or m^2).

Now round off this figure to something sensible as the shop will not be able to sell us 16.686 of a metre – call it 16.7m.

Having chosen a carpet, find the price – say it costs £5.36 per sq metre (sq m).

Now multiply 16.7 sq m by £5.36 per sq m (guess first).

$$16.7 \times 5.36 = £89.512$$

The answer is a sum of money, so round off to 2 decimal places to give £89.51, which will be the cost of the carpet.

Displaying and interpreting data

There are several ways of displaying data graphically, in charts or graphs. Look in any newspaper and you'll see examples of various graphs, tables and diagrams. It's useful to be able to interpret data displayed in various ways. Collect different types of graph and try to understand what they mean and why the information has been displayed in a particular way.

These examples are all based on the same data, as shown in the following table.

The figures represent the levels of production of goods (in thousands) by a company operating in three areas, East, West and North, over the four quarters of a year. For example, 20.4 thousand items were produced in the East region in quarter one, 27.4 in quarter two, and so on.

A quarter is three months. The first quarter starts from the beginning of the financial year. Different organisations operate with different financial years, for example, 1 April to 31 March, 1 August to 31 July, or 1 January to 31 December.

Financial data and data on production or consumption of goods are often displayed in this way.

	1st Quarter	2nd Quarter	3rd Quarter	4th Quarter
East	20.4	27.4	90	20.4
West	30.6	38.6	34.6	31.6
North	45.9	46.9	45	43.9

Table: Comparative production of goods (in thousands) for areas East, West and North.

Pie charts

A pie chart is a graph drawn as a circle and divided into pieces. Each piece is proportional to the size of the related piece of information, expressed as a fraction or percentage of the whole. Together the pieces make up a whole.

This pie chart shows the levels of production in the East area in all four quarters.

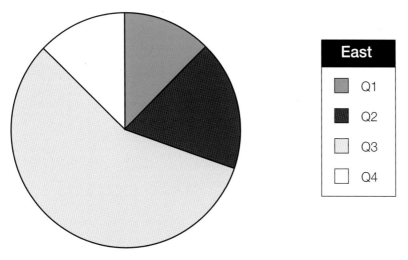

East
- Q1
- Q2
- Q3
- Q4

Bar charts

Bar charts, also known as column charts, consist of the two axes and a series of bars or columns. Each bar shows the value of each piece of information aligned with a scale. The bars are coded to represent different sets of information.

This bar chart shows the levels of production in each quarter in each of the three areas.

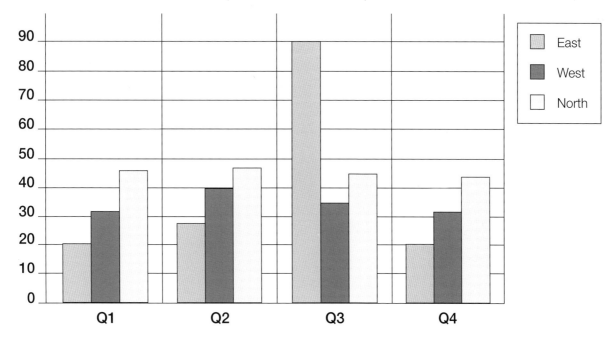

HD1/E3.1

Line graphs

Line graphs are a way of showing the relationship between two sets of information and how they vary on the two scales or axes. Several line graphs can be plotted on the same chart to visualise the relationships between sets of information.

The line graph, below, shows the data for each area in each quarter.

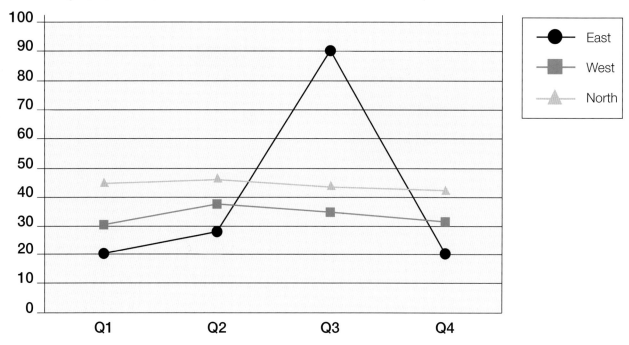

Area graph

An area graph is an alternative way of representing a line graph, highlighting the relative size of each of the sets of information.

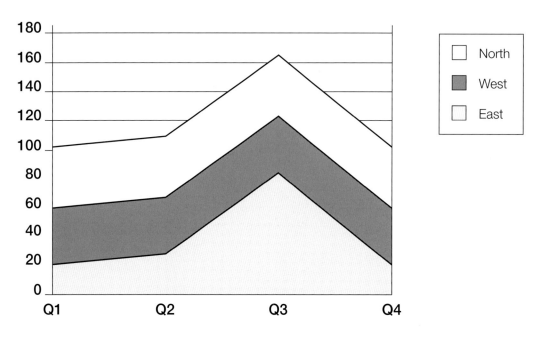

Checking your progress: mixed questions, p.3

1. £1.44
2. 136
3. 28 = 2 × 2 × 7
4. 18
5. £85
6. 53p
7. 25, 9
8. $\frac{5}{6}$
9. Write these as decimals, fractions and percentages: a half is 0.5, $\frac{1}{2}$, 50%; a quarter is 0.25, $\frac{1}{4}$, 25%; a whole one is 1, $\frac{1}{1}$, 100%.
10. Yes. The ratio 5:3 is the same as the ratio 15:9 because the relationship between the terms is the same. 5.3 is 15.9 expressed in its lowest terms.
11. 45
12. 1440 minutes
13. 89.54 or 89 remainder 6
14. 53p each
15. 12,792
16. £76
17. 12°C
18. £1.73
19. 8
20. 16

Puzzles, p.4

1. The traveller broke the chain like this:

   The first night he gave the landlord link 1.
   The second night he gave the landlord links 2 and 3 and got link 1 back in change.
   The third night he gave the landlord link 1 again.
   The fourth night he gave the landlord links 4, 5, 6 and 7 and got 1, 2 and 3 back in change.
   The fifth night he gave him link 1 again.
   The sixth night he gave the landlord links 2 and 3 and got link 1 back.
   The seventh night he gave him link 1.

2. 3 socks.

3. Ask, 'If I were to ask your brother which way is it to safety, what would he say?' Then take the opposite way.

Cross your numbers (1), p.5

¹3	4	²9	■	³5	⁴6
3	■	⁵3	⁶5	1	4
⁷9	2	6	7	■	■
0	■	⁸7	3	⁹8	0
¹⁰2	¹¹1	3	■	2	■
¹²8	1	6	5	3	0

Cross your numbers (2), p.6

¹3	²6	■	³5	2	⁴4	■	⁵6	0	⁶6	0
■	⁷2	0	0	■	7	■	2	■	2	■
⁸6	4	■	⁹3	¹⁰1	5	¹¹6	7	■	¹²4	¹³9
¹⁴1	4	¹⁵4	1	■	0	■	■	¹⁶2	5	6
5	■	¹⁷3	¹⁸1	2	1	5	■	9	■	2
■	¹⁹7	■	0	■	■	²⁰2	3	8	²¹1	■
²²4	2	■	²³3	5	²⁴1	■	■	²⁵3	2	²⁶4
²⁷4	0	8	8	■	²⁸8	6	6	4	■	0

Talking about time, p.7

120 minutes is 2 hours.
180 minutes is 3 hours.
360 minutes is 6 hours.
2 hours 25 minutes is 145 minutes.
There are 300 seconds in 5 minutes.
There are 48 hours in 2 days.

Answers

Write these times in figures, p.11

1. 6.00 or 18.00
2. 4.10 or 16.10
3. 7.25 or 19.25
4. 8.30 or 20.30
5. 8.15 or 20.15
6. 5.30pm or 17.30
7. 11.15am or 11.15
8. 10.25pm or 22.25
9. 2.05am or 02.05
10. 9.00 or 21.00
11. 4.40 or 16.40
12. 7.45pm or 19.45
13. 5.35 or 17.35
14. 3.45pm or 15.45
15. 2.55 or 14.55
16. 3.47 or 15.47
17. 9.26 or 21.26
18. 5.45 or 17.45
19. 1.17 or 13.17
20. 11.55 or 23.55

Time sense, p.12

11. Why do we call a century the 20th or 21st, although we say 1997, 2002 and so on? This is because years in the first century were numbered 1, 2, 3, and so on; those in the second century were numbered 101, etc., through to 1997 in the 20th century and 2002 in the 21st century.

The 24-hour clock, p.12

How would these times be written?
8.20 pm = 20.20
4.15 pm = 16.15
9.40 pm = 21.40
3.30 am = 03.30
6.30 pm = 18.30
7.00 am = 07.00
10.10 am = 10.10
5.45 am = 05.45
2.25 pm = 14.25

The digital clock, p.13

Write these times as they would be shown on a digital clock:
3 pm: 15.00
20 past 6: 06.20 or 18.20
half past seven: 07.30 or 19.30
10 to eight: 07.50 or 19.50
quarter to nine: 08.45 or 20.45
quarter past eleven: 11.15 or 23.15
25 to ten: 09.35 or 21.35
17 minutes past 4: 04.17 or 16.17
13 minutes to 5: 04.47 or 16.47
21 minutes to 1: 12.39 or 00.39

Reading a timetable, p.15

1. When does the 08.30 from Swansea arrive at Cardiff central station? 09.25
2. When does the 09.49 train from Port Talbot Parkway arrive in Bridgend? 10.00
3. I want to get to Oxford at about nine in the morning, to catch a connecting train. What train should I catch from Hereford to be there on time? 07.03
4. What time is the latest train you could take from Newport to get to London Paddington at around 11 am? 09.09
5. Your friend writes and asks you to meet her at Didcot Parkway off a train from Weston Super Mare. Her train leaves Weston at 12.15, what time does it get in to Didcot? 13.39
6. Does the 11.40 from Neath to Swindon stop at Bath Spa? No
7. Will I have to change trains if I catch the 11.55 from Cardiff Central on my way to Gatwick Airport? Yes
8. Can I get lunch on the 11.30 Swansea to Swindon train? No
9. My aunt said she'd meet me at London Paddington at 1.40 pm off the 12.35 train from Chippenham. I think she's misread the timetable, am I right? Yes
10. Is there a restaurant on the 10.30 train from Swansea? No
11. What do we mean by the 12.22 Bath Spa to London train? A train leaving Bath Spa at 12.22 which stops at London.
12. What time does the 13.10 from Bristol Temple Meads get in to Reading? 14.25

More time sense questions, p.16

1. 9.27pm
2. 16.18
3. 2 hours 41 minutes.
4. (work this out using a calendar)
5. 21 weeks and 5 days.

How long does it take?, p.17

1. 10 minutes
2. 3.52pm, 4 pm with the change made
3. 8.44 am
4. Tom:
   a. works 8 hours and 45 minutes altogether in the evenings
   b. he works 4 hours on Saturdays
   c. altogether he works 12 hours and 45 minutes each week
   d. he gets £70.13 for a week's work.
5. Jolene:
   a. works 26 hours and 15 minutes in the evenings during the week
   b. works 9 hours on a Saturday night
   c. earns £158.63 a week.
6. 44 minutes
7. 1 hour 30 minutes
8. 11.50 am
9. 1935
10. (take away 1986 from the current year)
11. Ben can he retire in 2017 when he's 65.
12. 13.30

Answers

The calendar, p.18

1. 12
2. 7
3. Spring, summer, autumn and winter
4. June
5. January
6. (today's date)
7. (the current season)
8. June-August is the hottest time of the year in Europe; December-February is the hottest time of year in Australia.

Writing the date, p.19

January 3rd 1999: 3.1.99
4th June 2003: 4.6.03
September 10th 1962: 10.9.62
May 21st 1942: 21.5.42
17 October 1951: 17.10.51
3.8.66: 3 August 1966
16.10.03: 16 October 2003
30/09/77: 30 September 1977
29.12.47: 29 December 1947
22.2.70: 22 February 1970
1/5/81: 1 May 1981

Days of the month – the calendar, p.20

1. There are 31 days in October.
2. There are 30 days in June.
3. There are 30 days in September.
4. There are usually 28 days in February.
5. There are 31 days in July.

The calendar – abbreviations, p.20

Jan is short for January
Feb is short for February
Mar is short for March
Apr is short for April
May is written in full as May
June is written in full as June
July is written in full as July
Aug is short for August
Sept is short for September
Oct is short for October
Nov is short for November
Dec is short for December

The days:

Mon is short for Monday
Tues is short for Tuesday
Weds is short for Wednesday
Thurs is short for Thursday
Fri is short for Friday
Sat is short for Saturday
Sun is short for Sunday

Ages, p.22

Fred is 19.
Julie will be 31 in seven years time.
Fred was 16 three years ago.
Julie will be 29 when Fred is 24.
What year was Fred born? (Take 19 away from the current year.)
The average age is 20.
The youngest was born 14 years ago (take 14 away from the current year.)
There are 11 years between the youngest and the eldest in Fred's family.

Pounds and pence, p.24

Change these sums of money to pounds:
55p is £0.55
88p is £0.88
101p is £1.01
172p is £1.72
2566p is £25.66
430p is £4.30

Giving change and getting change, p.25

1. 19p
2. £1.52
3. No, it should have been 17p
4. £1.00
5. £7.50

Going shopping, p.26

1. £4.48
2. 52p
3. 17.5.04

1. £7.14
2. £8.00
3. 86p
4. Yes

Shopping – find the cost, p.27

1. 1 kg of apples cost 90p, so:
   a. 2 kg of apples cost £1.80
   b. 3 kg of apples cost £2.70
   c. $\frac{1}{2}$ kg apples cost 45p
   d. 1$\frac{1}{2}$ kg apples cost £1.35
2. 1 packet of butter costs £1.09, so:
   a. 2 packets of butter cost £2.18
   b. 4 packets of butter cost £4.36
3. If 1 packet of butter weighs 250g, how much do:
   a. 2 packets weigh 500g
   b. 4 packets weigh 1000g or 1 kg
4. 125g of paté costs 76p, so:
   a. 250g of paté costs £1.52
   b. half a kilo of paté costs £3.04
5. 100g of bacon costs 65p, so:
   a. 200g of bacon costs £1.30
   b. 400g of bacon costs £2.60
   c. 300g of bacon costs £1.95

Answers

How much does it cost? p.27
1. £1.65
2. £12.60
3. £22.25
4. £90
5. £1.40
6. £5.94
7. £5.28
8. £3.15
9. £7.08
10. £6.40
11. £11.50
12. £23.97
13. £3.15
14. £5.55
15. £6

How much change to you get? p.28
1. 38p
2. 46p
3. £12.08
4. 37p
5. £1.24
6. £3.77
7. £2.55
8. £3.46
9. 23p
10. 75p

How much does it cost (2), p.29
1. £181.50
2. £5.22
3. £100
4. £47.95
5. £107.96
6. £3.60
7. £4.80
8. £55.93
9. £6604
10. £12.60
11. £12.95
12. £20.80
13. £344
14. £112.80
15. Approximately 65 litres.
16. £10.01

Driving, p.32
1. 198 miles
2. £21.58 Total bill: £29.31
3. 99 miles; 65502
4. 3 hours 45 minutes; 52.8 mph

Problems, p.32
1. £1
2. No. I should have been charged £21.86
3. 4.35pm; 5.00pm
4. 5.6 glasses

The metric system, p.37
1. 9.14 metres in 10 yards
2. 11.02 pounds in 5 kilograms
3. 109.36 yards in 100 metres

Units of measurement, p.38
1. There are 100 centimetres in one metre:
   100 cm = 1 m
2. There are 100 centilitres in one litre:
   100 cl = 1 l
3. There are 1000 metres in one kilometre:
   1000 m = 1 km
4. There are 1000 grams in one kilogram:
   1000 g = 1 kg
5. There are 1000 millimetres in one metre:
   1000 mm = 1 m
6. There are 10 millimetres in one centimetre:
   10 mm = 1 cm
7. There are 1000 millilitres in one litre: 1000 ml = 1 l

Weights and measures (imperial), p.39
1. There are 16 ounces in 1 lb.
2. There are 12 inches in 1 foot.
3. There are 20 fluid ounces in 1 pint.
4. There are 3 feet in 1 yard.
5. There are 36 inches in 1 yard.
6. There are 14 lbs in 1 stone.
7. There are 1760 yards in 1 mile.
8. There are 8 pints in 1 gallon.

Weights and measures (metric), p.40
1. There are 1000 grams in 1 kilogram.
2. There are 100 centimetres in 1 metre.
3. There are 1000 millilitres in 1 litre.
4. There are 1000 metres in 1 kilometre.
5. There are 1000 milligrams in 1 gram.
6. There are 1000 litres in 1 kilolitre.
7. There are 1000 millimetres in 1 metre.
8. There are 100 centilitres in 1 litre.
9. There are 10 decilitres in 1 litre.
10. There are 1000 kilograms in 1 metric ton.

Using the calculator to solve problems, p.48
£12.50 + £306.20 = £318.70
48.02 × 13 = 624.26
£46.23 − £13.95 = £32.28
11.65 × 7 = 81.55
£60.00 × 8 = £480
98.03 + 16.5 = 114.53
£96 ÷ 12 = £8
12.01 ÷ 3.2 = 3.753125
15 × £28.06 = £420.90
£45.02 + £19.36 = £64.38
£2.50 − £0.75 = £1.75
£890.07 + £96.90 = £986.97